Dominance Behavior

Jorge A. Colombo

Dominance Behavior

An Evolutive and Comparative Perspective

Springer

Jorge A. Colombo
National Scientific and Technical Research Council (CONICET)
Buenos Aires, Argentina

ISBN 978-3-030-97400-8 ISBN 978-3-030-97401-5 (eBook)
https://doi.org/10.1007/978-3-030-97401-5

This Springer imprint is published by the registered company Springer Nature Switzerland AG
The registered company address is: Gewerbestrasse 11, 6330 Cham, Switzerland

With love,
to my wife, Beatriz
to the memory of my parents and my brother, Emilio,
to my children and grandchildren.
to the memory of my mentors and friends Prof. Drs. Enrique Segura (Argentina), Charles Sawyer (USA), and Karl Zilles (Germany).

Preface

Our species has shown us its dark side and thrown shadows over its bright profile through the ages. The obscure pages of our civilization's history continue to be written by wars, conquest, dominance and oppression, torture, and slavery. These events occur under diverse forms—regional wars, modern forms of slavery, death and torture due to political or religious fanaticism, poverty and marginalization, gaps in cognitive development deepening the inequalities in social contexts, and degradation of our ecosystem for profit. This book explores some of the avenues that converge on the construction and consequences of our species' actions. Among them, the opposing criteria on the concept of culture and inequalities among social animals linked to our evolutionary backbone—i.e., the so-called behavioral universals in the animal kingdom.

In attempting to explore comparative events and delving into our species' bipolar nature and its consequences, this book provides elements of its complex intrinsic nature, human historical derangement, and perspectives of our social construction in the midst of opposing drives, such as dominance/predation/greed/privilege and solidarity/sharing/creativity. These tense our worldwide convivence and the construction of societies with equal rights within their cultural diversity. Our future path is at stake. The following excerpts provide a briefing of the conceptual basis that inspired this book along these main avenues.

> Obviously, if culture is defined as human culture, then only humans have culture.
>
> ...we believe that the study of culture would benefit from a comparative perspective, and that future work should address the question of whether various forms of culture are best viewed as falling along a continuum or as discrete categories. (Kuczaj and Highfill 2005).
>
> The study of inequality is essentially a concern with the evolution of human society and in fact is a predominant issue in recent considerations of social evolution.
>
> Human society operates within this didactic tension between dominance and equality, between hierarchical and egalitarian, between modes of behavior that feature or privilege the group to those that accent individuals. (Price and Feinman 2010).

Differences in social status among animals arise during competition for resources as winners display assertive behaviors and are recognized as dominant, while losers display submissive behaviors and are recognized as subordinate (Wilson, 1975; Herberholz et al. 2007). These behavioral differences result from changes in the nervous system induced by social experience. Social status affects neurogenesis in rodents (Kozorovitskiy and Gould, 2004) and crayfish (Song et al. 2007), neuronal size in fish (White et al. 2002), brain morphology in wasps (O'Donnell et al. 2007) and naked mole rats (Holmes et al. 2007), and cell receptor populations in crayfish (Spitzer et al. 2005) and fish (Burmeister et al. 2007). ... The variety and similarity of differences in the behavior of dominant and subordinate animals of many social species suggests that neural circuit reconfiguration that follows a change in social status is widespread among higher animals. (Issa et al. 2012).

Acknowledgment

To Flavia C. Abdullah, for language editing.

Contents

Chapter 1
Introduction

> We have not addressed the evolution of human social behavior here, but parallels with the scenarios of animal eusocial evolution exist, and they are, we believe, well worth examining. (Nowak et al. 2010)

> Heinlein (1973, p. 24) wrote *Never try to teach a pig to sing. It wastes your time and annoys the pig*, capturing succinctly a fallacy of much of comparative research. That is, as humans, we are often determined to test nonhumans for the ability to perform some task or display some feat that would demonstrate that they share a capacity known to be within the range of human abilities. This human-centered approach has been deemed the ´holy grail` approach to comparative research and poses serious problems for the study of animal behavior (Povinelli and Vonk 2003, 2004). (Shackelford and von Friedeburg 2012) (Italics added by JAC)

We study and care for what remains of our wildlife; we say we care about ecology, yet we abuse Earth's environment and risk generating critical environmental and climate events. We declare peace, and yet we inflict war and continue to develop deadly weapons or massive destruction and include projects that imply extraterrestrial power advantage. We forcefully and heatedly express our concern about poverty and marginality and our claims for equal rights. Yet, we produce conditions that oppose them and allow to generate elitist classes—that function as *castes*—with shameful power, financial privileges, offensive luxuries, and opulent richness. Common people are delegating on political powers crucial survival decisions for present and future generations.

Our species generates a world of contrasting events and inequalities in the collective and individual development of opportunities that nullify any future survival of human values and defeat any acceptable humanly based horizons. The very core of our human community has developed a sort of implicit schizoid conflict between opposing drives that menace our species and ecological survival. The power elites keep eroding our planetary ecology and building Star Wars-like dreams to increase profits from terrestrial and extraterrestrial resources, as well as competitive power stages. Financial and power caste systems keep deciding and manipulating our future. Financial and communicational power-driven elites and the generation of profound cognitive inequalities are deforming the very heart of our democracies and driving communities to profound cognitive gaps of unknown and unwarranted

J. A. Colombo, *Dominance Behavior*, https://doi.org/10.1007/978-3-030-97401-5_1

future conditions. In parallel with these dreadful developments, our species continuously show evidence of its creative power and people's solidarity.

The present essay approaches two fronts involved with human nature and actions. The first one tackles ancestral evolutive animal inertias that affect the basic construction of our brain/mind and social-behavioral spectrum. These proceed underneath cultural and political enclosures and represent basic universal behavioral components. The second front is a consequence of the first condition, i.e., the concept that basic, ancient, behavioral drives complicate the expression of a genuinely human democratic, equal rights' social construction. Those drives—that tint or condition our behavior—consist in *dominance/prevalence* (in all forms, social, political, financial, religious, reproductive, nutritional), *survival*, and, most of human origin, *social detachment* of elite minorities, financially powerful and drivers of human evolutionary trends, from general concerns and collective needs of legions of subdued populations. The quest for knowledge is embedded in our behavioral construction. Yet, its practical outcomes are significantly employed by opportunistic—political—strategies that seek dominance/prevalence, based on diverse degrees of profit that feed back on their factual and political power standing.

Besides the comparative complex richness and creativity of human culture in its various domains, the present essay focuses on the *basic behavioral drives* that would constitute a universal *ground-block* of animal strategies for endurance and survival, both on an individual and collective domain, regardless of its cultural complexity. Basic, ancestral, animal drives (*the universals*) lie beneath our sociocultural expressions and feed our most sophisticated expressions of survival, fight, flight, dominance, prevalence, submissiveness, cooperativity, and creativity. The most daunting expression of dominant behavior is met by our *human species* when it reverts into pathologically cruel behavior, as expressed in the following paragraphs (Colombo 2021):

> …the expression of predatory behavior surges periodically in the form of fundamentalisms, cultural prevalence, and belligerence, or arms race competitions, financial wars, religious wars, accumulation of material richness, and social dominance that often cross the borders into what could be considered as a cruel or insane domain. Pathological behaviors frequently emerged in the form of remarkable cruelty =a species-specific psychiatric twist of belligerent and aggressive components directed to silence critical voices, or to obtain information, to force bowing to human-made gods or to show personal power. The development of instruments of torture accompanied those behaviors, directed to silence critical voices or to obtain information, to force bowing down to human-made gods…
>
> This *specifically human capacity for deviant behavior* would have developed,as a backpack of our *sapiens* brain/mind potential.
>
> However, there are other forms for distorting the human mind and body that are exerted subtly, yet with a significant effect… that have succeeded to last to the present. These condemn people to degrading living conditions, identity loss… converting a potential creative citizen into a marginal being.

In addition to this basic drive of dominance that molds individual and social behavior, humans have construed or expressed additional profiles—whether of cognitive or emotional nature—that provide constructive means. These allow to expect the development of socially dynamic formats and the expression of individually creative or productive behaviors. Whenever these processes collide or collapse—i.e., dominance and creativity, dominance and self-assertion, dominance and basic freedom rights—individual and collective standings have generated social changes throughout history, open confrontations for predominance, or individual and collective (social) pathologies.

Bolhuis and Wynne (2009) stated that:

> As long as researchers focus on identifying human-like behaviour in other animals, the job of classifying the cognition of different species will be forever tied up in tickets of arbitrary nomenclature that will not advance our understanding of the mechanisms of cognition. For comparative psychology to progress, *we must study animal and human minds empirically, without naive evolutionary presuppositions.*

The emphasis is placed at exploring evidence of shared evolutive basic behaviors and drives observed across the animal kingdom, an ancient hereditary biological platform construed on universal basic needs, on which our species has mounted species-specific developmental cultural profiles.

As an *argument primer*:

> Even humans, now spread around the world and occupying habitats previously inaccessible to any primate, bear the genetic legacy of past population crises. All other branches of the genus Homo have passed into extinction. It may be that in the condition of Gorilla, Pan and Pongo we see some echo of our own ancestors before the last 100,000 years, and perhaps a condition experienced many times over several million years of evolution. It is notable that species within at least three of these genera continued to exchange genetic material long after separation[4,49], a disposition that may have aided their survival in the face of diminishing numbers. (Scally et al. 2012)

Which and how much of our current behaviors—both individually and as a global community—are driven by ancestral, inherited behavioral traits imprinted in our primal animal condition?

Throughout millennia, human civilization in its various geographic settings has developed different social structures. Perhaps, besides hunter-gatherers, a common feature of all of them was the systematic presence of commoners, or providers of services and goods, and social marginals; their lives subjected to rules and cultural conditioning. This was determined by forms of dominance/prevalence, of ruling powers, whether of authoritarian or pseudo representative structures. In modern times, they are enforced either by military, financial, and/or mass media communication means.

Interestingly, comparative examples surge in considering these structural social conditions, as if they would represent—on evolutionary grounds—ancient bridges with universals of social structures across evolutively distant species. Two components appear as constant motivators: survival drive (whether on physical terms or relative social standings) and dominance/prevalence.

Despite formal differences among species on how both domains—survival and dominance/prevalence—occur, social stratification remains a universal event adopting different forms, perhaps buffered in hunter-gatherer human groups. This enforced stratification could promote quasi-stable societies, as with eusocial species or human caste systems that evolved through civilization. In more open—"democratic"—conditions, the degree of stability would depend on the relative fitness of dominant and resistant groups.

Whatever the number of individuals eventually crossing the border in either direction, social stratification remained a persistent condition. Circumstantial evidence associated with such comparative evolutionary and modern human cultural developments will be considered. Behind the complexity of human sociocultural characteristics, signs and clues of shared basic motivations can be found across various animal species and the history of human civilizations. The installment of inequality within social species will be accessed on comparative and evolutive grounds.

Hence, the following account aims at providing possible clues supporting an evolutive concept of the apparent main behavioral domains in nature that build human ancestral backpack. These are dominance/prevalence, resistance, submission, learning, and inheritance. Across this virtual trajectory and behavioral expressions, it is proposed that—depending on the species' interactive behavioral complexity—a sequence of social power expressions could be envisioned as being represented by castes, classes, values, and prejudices. This assertion takes into consideration the variable components and profiles that result from genetic and cultural interactions expressed as social structures or cultural formats.

In this context, the concept of culture applies to species-specific codes, fixed behavioral components or traditions, and instruments that conform to a community, embedded as implicit or manifest forms of dominance/prevalence (or imperative), resistance (or dialectic, or competitive), submission (or acceptance, or cooperation), learning, and inheritance. Within this conceptual context of *culture,* it seems appropriate to include the following statements (Whiten et al. 2017):

> Recent decades have revealed that social learning and the transmission of cultural traditions are much more widespread in the animal kingdom than earlier suspected, affecting numerous forms of functional behavior and creating a secondary form of evolution, built onto the better-known primary, genetically based form.
>
> Social learning is now extensively documented in mammals (29), with a particular intensity of research studies in primates (30–33) and cetaceans (34–37), in birds (38–41), in fish (42), and in insects (43, 44).

An evolutive, comparative view of the development of sociality in humans following the (proposed) historical sequence of caste, race, class, and prejudice implies an analysis of the evolution of social species and their communal organization, including the concept of culture as a shared set of behavioral profiles, cues, and skills. As it will be analyzed, this amplified concept of culture implies avoiding a limited human view, with its rich sets of symbolism, creativity, values, and technological development.

As stated by Richerson (2011):

> Robert Boyd and I use the following psychologically-based working definition of culture: Culture is information capable of affecting individuals' behavior which they acquire from other members of their species through teaching, imitation, and other forms of social transmission. By this definition, culture has proven to be widespread in the animal kingdom.

Comparative analysis of individual and collective behaviors in social species faces a dilemma. It is adequately expressed by Kappeler et al. (2019), considering the extreme diversity among animal species (including humans):

> … it remains an open question whether it is possible to construct a single index of social complexity that can be applied across taxonomically diverse species or whether different indices of social complexity have to be used in different lineages. … it is still an open question whether social complexity is a property of a society that is measurable on a single scale and whether it is useful to attempt to compare the complexity of different societies.

The following analysis will focus on a series of reported comparative events and essential—individual and group—behavioral profiles. These share cues that link social evolution across the animal kingdom and would have collectively emerged into social components of the progressively complex human cultural and social diversity.

References

Bolhuis JJ, Wynne CDL. Can evolution explain how minds work? Nature. 2009;458:832–3.

Colombo JA. The Homo within the sapiens. New York: Nova Science Publishers; 2021.

Kappeler PM, Clutton-Brock T, Shultz S, Lukas D. Social complexity: patterns, processes, and evolution. Behav Ecol Sociobiol. 2019;73:1–5. https://doi.org/10.1007/s00265-018-2613-4.

Nowak MA, Tarnita CE, Wilson EO. The evolution of eusociality. Nature. 2010;466:1057–62. https://doi.org/10.1038/nature09205.

Richerson PJ. Cultural evolution. In: Yan-jie S, Wang XT, editors. Thus spake evolutionary psychologists. Beijing: Peking University Press; 2011.

Scally A, Dutheil JY, Hillier LW, Jordan GE, Goodhead I. Insights into hominid evolution from the gorilla genome sequence. Nature. 2012;483:169–75.

Shackelford TK, von Friedeburg R. Toward bridging gaps: finding commonality between evolutionary and comparative psychology. In: Shackelford TK, Vonk J, editors. The Oxford handbook of comparative evolutionary psychology. New York: Oxford University Press; 2012. https://doi.org/10.1093/oxfordhb/9780199738182.013.0001. www.pnas.org/cgi/doi/10.1073/pnas.1707630114.

Whiten A, Ayala FJ, Feldman MW, Laland KL. The extension of biology through culture. PNAS. 2017;114:7775–81. www.pnas.org/cgi/doi/10.1073/pnas.1707630114

Chapter 2
Eusociality

> We have not addressed the evolution of human social behavior here, but parallels with the scenarios of animal eusocial evolution exist, and they are, we believe, well worth examining. (Nowak et al. 2010)

> The evolution of eusociality in several insect lineages (e.g., ants, honeybees, vespid wasps, termites) provides a striking example of convergent phenotypic innovation. Eusocial insect societies are founded upon a novel caste polyphenism[1], in which reproductive queen and nonreproductive worker female castes develop from the same genome, depending mainly on socially regulated nutritional inputs. ...results emphasize that the recruitment of both highly conserved and lineage-specific genes underlie the convergent evolution of novel traits such as eusociality. (Warner et al. 2019)

> Reproductive division of labor is one of the defining characteristics of eusociality: queens reproduce, and their sterile worker daughters gain indirect fitness by helping to provision the queen's offspring, who are their sisters (Michener 1974). Physiological differences between queens and workers also extend to the brain (Molina and O'Donnell 2008, O'Donnell et al. 2007, 2017, Smith et al. 2010, Rehan et al. 2015). (Pahlke et al. 2019)

Eusociality, Definitions

> The evolutionary origins of advanced eusociality, one of the most complex forms of phenotypic plasticity in nature, have long been a focus within the field of sociobiology. (Shell and Rehan 2019)

According to the Encyclopedia Britannica:[2]

> Eusocial species, any colonial animal species that lives in multigenerational family groups in which the vast majority of individuals cooperate to aid relatively few (or even a single) reproductive group members. Eusocial species often exhibit extreme task specialization, which makes colonies potentially very efficient in gathering resources...

[1] The phenomenon where two or more distinct phenotypes are produced by the same genotype.

[2] https://www.britannica.com/science/eusocial-species

J. A. Colombo, *Dominance Behavior*, https://doi.org/10.1007/978-3-030-97401-5_2

Chapman et al. (2000) added:

> In eusocial species, the first cohort to eclose are gall-bound soldiers...

Eusocial insects provide multiple examples of genetic and environmental interactions in the construction of social caste systems. Although this shows a certain degree of plasticity in establishing social casts, its final structure is always characterized by hierarchical stratification. This consists of a queen and its controlled generation of sterile helpers and sterile workers. Hence reproductive potential becomes a crucial parameter in the establishment and development of social structure among insect species. According to Warner et al. (2019):

> The evolution of eusociality in several insect lineages (e.g., ants, honeybees, vespid wasps, and termites) provides a striking example of convergent phenotypic innovation[7]. Eusocial insect societies are founded upon a novel caste polyphenism, in which reproductive queen and nonreproductive worker female castes develop from the same genome, depending mainly on socially regulated nutritional inputs.[8, 9]

Weiner and Toth (2012) remark the extreme phenotypic genomic plasticity of social insect castes inasmuch as differences are environmentally (sensitive to the abiotic and social environment), not genetically, determined.

As discussed below, this stringent definition has been challenged by scientists in the field, which opens a broader conceptual framework. According to Nowak et al. (2010), eusociality is present in species:

> ... in which some individuals reduce their own lifetime reproductive potential to raise the offspring of others.

According to Brand and Chapuisat (2012):

> The hallmark of eusociality is reproductive division of labour between generations, a surprising social organization by which some individuals become functionally sterile helpers [1].

Although not representing a majority among insect species, as pointed out by Wilson and Holldobler (2005), eusociality:

> ... is consistent with a remarkable phenomenon that has been in plain sight but largely overlooked by investigators: the phylogenetic rarity of eusociality. Of the 2,600 living taxonomic families of insects and other arthropods currently recognized, only 15 are known to contain eusocial species (13, 24, 37).

It represents a notable social organization present in insects[3] and some mammals, as described by O'Riain et al. (2000) in the eusocial naked mole rat, as we will consider later. These authors inquire themselves whether the evolution of morphological castes is phylogenetically constrained to invertebrates. As mentioned above, close similarities in social structures are found among some human societies. This, to the point that Nowak et al. (2010) have stated that it:

---

[3] Predominant in Hymenoptera (insects with membranous wings), which represents a large order of insects, comprising the sawflies, wasps, bees, and ants. Over 150,000 living species of Hymenoptera have been described, in addition to over 2000 extinct ones. https://en.wikipedia.org/wiki/Hymenoptera

> ...underlies the most advanced forms of social organization and the ecologically dominant role of social insects and humans.

Eusociality introduces the role of castes, a concept that in human terms has been applied to specific societies in which all previous qualifiers meet and in the present essay is expanded to modern forms of political power delegation, that is, in which biological reproduction is replaced by power delegation that extends in time socio-political structures or privileges. This proposal expands the notion of its origin as claimed by Berreman (1967):

> A caste system occurs where a society is made up of birth-ascribed groups which are hierarchically ordered and culturally distinct. The hierarchy entails differential evaluation, rewards and association.

According to the Encyclopedia Britannica,[4] the concept of *caste* implies:

> ...any of the ranked, hereditary, endogamous social groups, often linked with occupation...

According to Crespi and Yanega (1995), the concept of *castes* would:

> .... have been first coined by Battra (1966) to describe nesting behavior in bees, where the nest founding individuals survive to cooperate with a group of their mature daughters with division of labor, a concept further qualified by Michener (1969) and Wilson (1971, 1975).

This further adds to their concept of caste:

> ... groups of individuals that become irreversibly behaviorally distinct at some point prior to reproductive maturity.

Shell and Rehan (2019) stress the phenotypic plasticity in the evolutionary emergence of insect sociality, a concept that would include a broader spectrum of species regarding the impact of social context on individual development. In this regard, consideration includes the impact of social environment on humans (reviewed as epigenetic factors in Colombo 2019 and developed as a research topic involving poverty and brain and cognitive development as reviewed in Colombo 2007 and Lipina and Colombo 2009).

Eusociality, Conceptual Expansion

> Cooperative breeding and eusociality are not discrete phenomena, but rather form a continuum of fundamentally similar social systems whose main differences lie in the distribution of lifetime reproductive success among group members. Sherman et al. (1994)

According to Lacey and Sherman (2005), the field is traversed by a semantic debate on the definition of eusociality. As mentioned, the term was coined by Batra in 1966

[4] https://www.britannica.com/topic/caste-social-differentiation

(cf. Lacey and Sherman 2005), based on the following criteria obtained from bee's behavior: living in multigenerational groups, alloparent caring of young, and restricted reproduction. This term was expanded to include several orders and species in which those three criteria were met (Sherman et al., 1994) and considered part of a eusociality continuum. According to Bourke (2011):

> Social evolution is a central topic in evolutionary biology, with the evolution of eusociality (societies with altruistic, non-reproductive helpers) representing a long-standing evolutionary conundrum. Recent critiques have questioned the validity of the leading theory for explaining social evolution and eusociality, namely inclusive fitness (kin selection) theory.

Also, Bourke states that the formal version of inclusive fitness theory as summarized by Hamilton's rule (1961) (cf. Bourke 2011) shows that, other things being equal, individuals should behave toward others as if they valued their reproduction in proportion to how related they are. This position is argued against by Nowak et al. (2010) based on the standing that fitness theory provides no additional insights to those provided by modeling social evolution population genetic approaches and an alternative theory for eusocial evolution.

Canciani et al. (2019) stated that eusociality was at the center of many debates in philosophy and biology for decades because it represents an extremely high form of social integration. It should be added that it involves an organization with a hierarchical sustain of roles based on selective early feeding (food and cognitive menus), that is, a process that involves a condition of early, initial, developmental inequality from which fixed community roles are imposed. Neural and behavioral plasticity is a conserved resource in case of disruptive events. Unless we take into consideration these basic tenets, extrapolations to other species may involve misleading arguments. According to Ratnieks and Helantera (2009), eusociality represents a social system that involves coercion and led to inequality, thus conceptually placing it close to known sociopolitical systems and populations in marginal conditions in human civilization. Their interesting view approximates several social conditions found in human societies:

> Coercion has evolved after eusociality and acts to prevent individuals attempting to reproduce instead of working. Without coercion, more insect societies would be like colonies of Melipona or queenless honeybees, in which a large proportion of colony resources and individual lives are directed into intra-colony competition over reproduction rather than into working to increase the colony's total reproduction. Insect societies are not the only type of social group in which the interests of different individuals vary and in which social coercion plays a role in reducing conflict (West et al. submitted; Ratnieks and Wenseleers 2008). In insect societies, this has led to extreme inequality.

Kapheim et al. (2015) point out that relatedness is an essential component of eusocial evolution in which helpers do not necessarily improve their inclusive fitness by foregoing reproduction, and its early evolution was favored by maternal manipulation. Interestingly, if we extrapolate this concept to human behavior, parental figures (such as motherland or paternal leadership) do have an individual and collective impact on social dynamics, as will be reviewed later. Furthermore, at the human level, ancient time slavery and modern expressions of it have the social construction profile and group effects that resemble forms of imposed eusociality.

Wilson and Holldobler (2005) consider that eusociality in ants and termites in the irreversible stage is the key to their ecological success. According to these authors, their colonies are consistently superior to solitary and pre-eusocial competitors due to the altruistic behavior among nestmates and their ability to organize coordinated action by pheromonal communication. These authors define eusociality as colonies consisting of overlapping generations, cooperative brood care, and a reproductive division of labor where sterile (or nonreproductive) workers help the reproductive members.

Nowak et al. (2010) consider eusociality where adult members are divided into reproductive and (partially) nonreproductive castes and the latter care for the young. Within this concept, they *loosely* include humans as eusocial, although this would include the concept of caste over the notion of kin. These authors state that:

> Eusociality, in which some individuals reduce their own lifetime reproductive potential to raise the offspring of others, underlies the most advanced forms of social organization and the ecologically dominant role of social insects and humans.

This expanded concept of eusociality, though avoiding including the concept of caste, is shared by Hardisty and Cassill (2010):

> Some researchers would narrow the definition of eusociality to exclude vertebrates (Crespi and Yanega 1995, contra Sherman et al. 1994; Burda et al. 2000; Foster and Ratnieks 2005). However, we propose that the definition be expanded to include social vertebrate families consisting of one or more reproductive females, their offspring and other non-reproductive helpers such as unmated aunts/uncles, grandparents and nannies that help rear the reproductive female's offspring.

Should definition include the formal application of the concept of *caste*, it would seem to contradict widely diverse current human trends in the cultural and sociopolitical organization and the quest for freedom of individual decision, group selection, and offspring care. Although applied to ancient and some modern societies, in others, the concept of *caste* and division of labor was reformulated in terms of modified social dimensions, such as formal and virtual inequality expressed in *class*, social *prejudice,* and social stratification involving poverty and marginalization. Sherman et al. (1994) considered that:

> Eusocial societies are traditionally characterized by a reproductive division of labor, an overlap of generations, and cooperative care of the breeders' young. Eusociality was once thought to occur only in termites, ants, and some bee and wasp species, but striking evolutionary convergences have recently become apparent between the societies of these insects and those of cooperatively breeding birds and mammals. These parallels have blurred distinctions between cooperative breeding and eusociality, leading to calls for either drastically restricting or expanding usage of these terms. We favor the latter approach. Cooperative breeding and eusociality are not discrete phenomena, but rather form a continuum of fundamentally similar social systems whose main differences lie in the distribution of lifetime reproductive success among group members.

According to Shell and Rehan (2019) and Shell et al. (2021), eusociality represents one of the most complex forms of phenotypic plasticity and driver of macroevolutionary genomic changes, according to theories stating that modular regulation and deeply conserved genes may play essential roles in both the evolutionary emergence and elaboration of insect sociality.

At variance with less restricted views, Scantlebury et al. (2006) propose that among mammalian species, eusociality would only take place in two species of African mole rats (Fig. 2.1). It is characterized by the division of labor between morphologically distinct "castes." Thus, reproductive division of labor is one of the defining characteristics of eusociality: queens reproduce, and their sterile worker daughters gain indirect fitness by helping to provide the queen's offspring. This social stratification would have an impact on a neural organization, since as mentioned by Pahlke et al. (2019):

> Physiological differences between queens and workers also extend to the brain (Molina and O'Donnell 2008, O'Donnell et al. 2007, 2017, Smith et al. 2010, Rehan et al. 2015).

This would occur as the consequence of differential feeding and later by the control by the queen ("oppression") as the primary favor leading to the appearance of a worker caste. Following Michener and Brothers (1974), worker altruism is a necessary consequence of eusocial systems.

From a contemporary human equal rights standpoint, it sounds contradictory to qualify caste-based social structures as being eusocial, based on attaining a fixed—or quasi-stable—role for each of its constituents, opposed to the concept of maximizing individual freedom within an organized community. From this perspective, the latter would seem more valuable than establishing fixed roles, a trend toward high specialization that we can perceive in modern times (this theme will be further developed later). Besides the plasticity involved in attaining such eusocial structure, as Shell and Rehan (2019) mentioned, it involves a stark division of labor and roles

Fig. 2.1 Naked mole rat. (Felix Petermann, MDC)

in the community. Some insect species express different options in terms of social organization.[5]

As stated by Powell and Franks (2005):

> The great evolutionary and ecological success of ant societies (Holldobler and Wilson 1990) can be attributed to the efficiencies that accrue from the division of labour within their colonies (Oster and Wilson 1978; Holldobler and Wilson 1990). This organizational system is underpinned by individual specialization, but paradoxically, morphologically distinct specialists within the general workforce are rare (Oster and Wilson 1978; Baroni Urbani 1998).

According to Tian and Zhou (2014):

> The presence of reproductively altruistic castes is one of the primary traits of the eusocial societies. Adaptation and regulation of the sterile caste, to a certain extent, drives the evolution of eusociality.

Although the term *eusocial*, as coined by Battra in 1966 (cf. Crespi and Yanega 1995), describe nesting behavior in bees and later formalized by Michener (1969; cf. Crespi and Yanega 1995) to comprise castes and division of labor, cooperative work, and generation overlap, this latter condition was not considered as necessary by Crespi and Yanega (1995). Based on this conceptual reference and those quoted previously (Sherman et al. 1994; Ratnieks and Helantera 2009; Nowak et al. 2010; Hardisty and Cassill 2010), let us proceed with the expanded eusocial concept of caste, thus considering the conceptual expansion that involves insects and mammalian species.

Furthermore, in considering evolutive representations in higher-order species, these behaviors, highly driven by genetic signals but conditioned by environmental (social and physical) factors, draw interesting behavioral outcome parallelisms with social organization in other species (to be dealt later in its possible comparative, evolutive implications for the human species). However, the definition of eusociality has generated different opinions—to include variant forms among species. One such position is claimed by Sherman et al. (1994) supporting a less narrow concept than the one based essentially on the reproductive division of labor and be applied only to a subset of insects currently recognized as eusocial:

> Alternatively, definitional problems could be reduced by expanding the eusociality concept to include all vertebrate and invertebrate societies with helpers. We favor the latter approach. It seems more productive to recognize that similar social systems occur in birds, mammals, and insects than to debate whether particular insects are eusocial (e.g., Furey, 1992 versus Tsuji, 1992).

[5] Besides some insect species such as primitively eusocial Halictidae (is the second largest group of bees), in which a low degree of differentiation occurs between queens and helpers, as mentioned by Brand and Chapuisat (2012): "Primitively eusocial halictids have a low degree of morphological differentiation between queens and helpers and a high degree of behavioural flexibility in both types of individuals [2]. As a result, females have multiple reproductive options that result in diverse types of social organisation."

This amplified concept provides grounds for evolutive, comparative considerations, as it will be reviewed later.

As stated by Alward et al. (2020):

> Social hierarchies are ubiquitous in social species and profoundly influence physiology and behavior.

Interestingly, Tian and Zhou (2014) noted species differences as to whether workers or soldiers represent the first evolved sterile cast:

> Depending on adaptive functions of the first evolved sterile caste, eusocial societies can be categorized into the worker-first and soldier-first lineages, respectively. The former is marked by a worker caste as the first evolved altruistic caste, whose primary function is housekeeping, and the latter is highlighted by a sterile soldier caste as the first evolved altruistic caste, whose task is predominantly colony defense. The apparent functional differences between these two fundamentally important castes suggest worker-first and soldier first eusociality are potentially driven by a suite of distinctively different factors. Current studies[3] have been focused largely on the worker-first Hymenoptera, whereas understanding of soldier first lineages including termites, eusocial aphids, gall dwelling thrips, and snapping shrimp, is greatly lacking.

Neurobiological and Genetic Components

Worker sterility develops either because of physical dominance (bees and wasps, primitive eusocial species) or mediated through the queen's pheromone signals (honeybees, advanced eusocial species). Besides phenotypic and physiological differences between queen and sterile workers observed in bees, size differences exist in a brain area related to sensory integration, learning, and memory in the so-called mushroom body (MB), a higher-order center for sensory integration and memory. According to Pahlke et al. (2019):

> The larger MBs of queens in primitively eusocial insects may be a plastic response to the social interactions involved in establishing dominance or result from developmental differences between queens and workers before a social colony is established.

As Brand and Chapuisat (2012) have stated:

> Primitively eusocial halictids[6] have a low degree of morphological differentiation between queens and helpers and a high degree of behavioural flexibility in both types of individuals [2]. As a result, females have multiple reproductive options that result in diverse types of social organisation. (Footnote[6] added by JAC)

Behavioral evolution among Hymenoptera[7] insects involved changes in the neural substrate of the mushroom body, as described by Oya et al. (2017), and comprised several species exhibiting various behavioral traits (solitary and phytophagous, parasitic).

---

[6] A species family of bees.

[7][7] A large order of insects with membranous wings that include the sawflies, wasps, bees, and ants.

The authors mentioned above found structural changes in the mushroom body related to neural evolution consisting of Kenyon cell subtypes (mushroom bodies' intrinsic neurons) becoming more diverse with the behavioral (parasitic lifestyle and nidification) evolution of Hymenoptera.

Phenotypic plasticity in alternative queens' and workers' castes in social insects, particularly honeybees, would depend on epigenetic (environmentally triggered) modifications of DNA (specifically, chemical modifications of proteins linked to DNA that do not change the DNA sequence) through methylation that could expand genomes' regulatory flexibility (Weiner and Toth, 2012).

According to Warner et al. (2019), following analysis of two species of insects that represent two independent origins of eusociality, these authors identified a substantial shared core of genes upregulated in the abdomens of queen ants and honeybees. These genes also tend to be upregulated in mated female flies, suggesting that they are part of a conserved insect reproductive ground plan.

These studies converge with those of Shell and Rehan (2019), as they report that:

> The evolutionary origins of advanced eusociality, one of the most complex forms of phenotypic plasticity in nature, have long been a focus within the field of sociobiology. Although eusocial insects are known to have evolved from solitary ancestors, sociogenic research among incipiently social taxa has only recently provided empirical evidence supporting theories that modular regulation and deeply conserved genes may play important roles in both the evolutionary emergence and elaboration of insect sociality.

However, the expression of these deeply conserved genes is subjected to the effects of differential feeding during the larval period, as described in Barchuk et al. (2007) and Moda et al. (2013).

As mentioned earlier, in eusocial insects, the production of daughters is generally restricted to mated queens, and unmated workers are functionally sterile. In honeybees, the reproduction of workers is regulated by the queen, brood pheromones, and workers' control. However, some workers can escape from this and establish themselves as social parasites by activating their ovaries and producing queenlike amounts of queen pheromones influenced by a locus in chromosomes (Jarosch et al. 2011). As mentioned by Moritz et al. (2011), this parasitic component in insect social structure (described in honeybees) occasionally plays a role in queen replacement. These authors sustain that hopelessly queenless colonies typically die due to parasitic takeovers and the parasitic laying workers are an essential life-history strategy more frequently used than in providing a native queen to rescue the colony.

Regarding phenotypic *plasticity* in insect caste systems, Weiner and Toth (2012) remark that the extreme phenotypic plasticity of social insect castes has become even more compelling with the knowledge that, with some notable exceptions (ants genera), in most species, queen and worker caste differences are environmental and not genetically determined.

> Evidence to date suggests the effects of diet on caste phenotype can be mediated by methylation of particular genes [21] ... In our conceptual framework ... we propose that dietary differences lead to differential methylation. This, in turn, leads to alternative splicing and possibly caste-biased expression, which leads to caste-biased phenotypes, such as restricted ovarian development in workers or larger body size and longer lifespan in queens... (Weiner and Toth, 2012)

Genetics and molecular research have added possible mechanisms for the stability and plasticity of the caste system, at least as it could be applied to some insect species. According to Lattorff et al. (2007), caste determination in bees occurs at an early stage of larval development and is independent of the genotype but depends on differential nutrition, while workers refrain from reproduction. However, Cape honeybee workers can produce workers by parthenogenesis—as determined by a single gene—that develop queenlike phenotypes in their physiology and behavior. Moda et al. (2013) presented a list of genes differentially expressed between queens' and workers' brains induced by differential feeding during the larval stage.

Social and Environmental Contexts and Brain Development. Eusocial Mammals

> Social rank along a hierarchy determines physiological state and behavioral performance. A ubiquitous feature of social hierarchies is the communication of rank through nonphysical signaling systems (e.g., coloration) and aggression, traits that correlate with the reproductive status of an individual. Social hierarchies are ubiquitous in social species and profoundly influence physiology and behavior. (Alward et al. 2020)

It should be stressed, though, that some researchers would narrow the definition of eusociality, a debatable field according to Hardisty and Cassill (2010):

> ...we propose that the definition be expanded to include social vertebrate families consisting of one or more reproductive females, their offspring and other non-reproductive helpers such as unmated aunts/uncles, grandparents and nannies that help rear the reproductive female's offspring. (Hardisty and Cassill 2010)

Besides insects, these general statements not only involve nonmammalian vertebrates, as stated by Goodson (2005):

> Based on a wide variety of data, it is now clear that birds and teleost (bony) fish possess a core 'social behavior network' within the basal forebrain and midbrain that is homologous to the social behavior network of mammals. (Goodson, 1995)

However, according to Goodson (2005) and first described for mammals by Newman (1999), the social behavior network is a fundamental and evolutionary conserved feature of the vertebrate brain.

> In what way could a common neuroanatomical network provide a substrate for the broad repertoire we must consider in the category of social behaviors? Again, borrowing an important concept from our colleagues in cognitive neuroscience, we envision that a particular social behavior, for example, male sexual behavior, is an emergent property of the pattern of activity across the network[95]. (Newman, 1999)

This includes eusocial mammals as the naked mole rat, a species with the most rigid organized reproductive hierarchy among mammals, according to Holmes et al. (2007). In this eusocial species, a change in social status triggers neural remodeling as a predominant role over sex.

Brain remodeling would affect the ventromedial hypothalamic nucleus, bed nucleus of the stria terminalis, paraventricular nucleus, and amygdala. This impact of social status on the brain represents a significant antecedent for future considerations on the impact of sociality on brain/mental development in higher-order mammalian species, including humans, as stated by Nowak (2010):

> Eusociality, in which some individuals reduce their own lifetime reproductive potential to raise the offspring of others, underlies the most advanced forms of social organization and the ecologically dominant role of social insects *and humans*. (Italics by JAC)

Analysis of neuronal numbers and proliferating neurons in the hippocampus of three species of mole rats showed higher numbers of young neurons in the Highveld and naked mole rats, as compared with solitary Cape mole rats, suggesting that neuronal differentiation in these eusocial species reflects species-specific demands (Amrein et al. 2014). Another variable should be included in these correlations, which relates to social status in terms of dominance or subordinate condition (Kozorovitskiy and Gould 2004). Such roles establish nexus with various forms of social structures observed in higher-order mammalian species, including humans, as we shall consider later, with the inclusion that socially skewed labor access—inequalities—also results in altruism.

Interestingly, social hierarchies impact the genetic expression leading to brain development, as observed in honeybees' castes, in which early differential feeding regimes generate a complex cascade leading to differences in brain morphologies between queen and workers (Moda et al. 2013). Queen phenotype is driven by epigenetic mechanisms of gene regulation, mediated by differential DNA chromatin methylation, according to Vaiserman (2014) and Wojciechowski et al. (2018). In primitive eusocial insects, a brain area called the mushroom body associated with sensory integration and learning is larger in queens than in workers, which may be associated with increased nutrition access during the larval stage (Pahlke et al. 2019).

Barker et al. (2021) used playback experiments to demonstrate that individuals make preferential vocal responses to home colony dialects, which decrease and remerge with queen loss and its replacement, suggesting the queen partially controls it. According to these authors:

> Pups fostered in foreign colonies in early postnatal life learn the vocal dialect of their adoptive colonies, which suggests vertical transmission and flexibility of vocal signatures.

At the early stages of development, this dependence on nutrition quality and access bears an interesting close relationship with conditions imposed on humans by extreme poverty and marginal social conditions.

Naked mole rats perhaps express the most typical caste social structure among mammalian vertebrates. Several authors (e.g., Jarvis and Sherman 2002) described that in naked mole rat colonies, reproduction is restricted to one female, and usually up to two males and nonbreeders are not physiologically sterile. However, they are sociologically suppressed by the aggressive dominance of the breeding female. In naked mole rat societies, hierarchies are based on size-based dominance, in which breeders' largest individuals of both sexes obtain access to food or digging sites before smaller individuals.

Recent observations on the naked mole rat behavioral organization added a new element in the concept of culture within the animal kingdom. Barker et al. (2021) showed that calls ("chirps") emitted by individuals convey information specific to the animal's group, thus providing the basis for cultural clues among this species. According to these authors, the most common naked mole rat vocalization, the soft chirp, is used to transmit information about group membership, creating distinctive colony dialects, thus providing a means for their highly organized social structure. Colonies of naked mole rats develop dialects in their vocalizations that may help them distinguish between friends and foes. These dialects are influenced by each colony's queen and become more varied if the queen dies. This species can grow colonies of up to 200 individuals, a feat that implies group and individual identity and pertinence, based on dialect features that can be transmitted across generations, thus, according to the authors, supporting the concept that social complexity evolved concurrently with vocal complexity and plasticity since individual vocal patterns in this species change with development under different queens.

Hierarchical access to feeding and housing represents privileges also expressed in higher-order mammals and cultured-modified privileges among human societies. Kverkova et al. (2018) discuss possible factors linking social complexity and brain size in species with monopolized reproduction, such as the naked mole rat, among them, social bonding, cooperation, and Machiavellian behavior.

According to Perez-Barberia et al. (2017), sociality and brain size can become decoupled in some mammalian species. Regarding neuronal replacement during adulthood, Gheusi et al. (2009) pose that it is widely distributed across species and is probably common to all vertebrates.

Concerning humans, Lahdenpera et al. (2004) and Foster and Ratnieks (2005) refer to middle-aged women that become sterile via menopause and thus able to help the next generation (the "mother" and "grandmother" effect):

> Eusociality, in which some individuals reduce their own lifetime reproductive potential to raise the offspring of others, underlies the most advanced forms of social organization and the ecologically dominant role of social insects and humans. Most animals reproduce until they die, but in humans, females can survive long after ceasing reproduction. In theory, a prolonged post-reproductive lifespan will evolve when females can gain greater fitness by increasing the success of their offspring than by continuing to breed themselves. (Lahdenpera et al. 2004)

> Recent data show that distinct reproductive and helping strategies also occur in a more familiar vertebrate species, but in the reverse temporal sequence to mole rats. In mid adult life, half the breeders become physiologically incapable of reproducing and help their close relatives. *Uniquely for a vertebrate, the helpers are permanently sterile. What species is this? It is our own.* (Foster and Ratnieks 2005) (italics from JAC)

This concept raised a warning by Burkart et al. (2017), establishing a different origin of cooperative breeding in humans, an argument held by various authors suggesting the potential heterogeneity of cooperative offspring care in different species or lineages. This statement does not invalid similarities or parallelisms suggestive of common, basic, ancestral drives in animal behavior (see later)—though its expression is adapted to different socioecological scenarios—as mentioned by Nowak et al. (2010) regarding humans:

> We have not addressed the evolution of human social behavior here, but parallels with the scenarios of animal eusocial evolution exist, and they are, we believe, well worth examining.

Concerning the developmental timing for the expression of morphological divergence, according to O'Riain et al. (2000), differences exist between eusocial insects and mammals:

> In all of the ants and a proportion of the wasps and bees (including *Apis*), adult females exhibit two distinct phenotypes as a consequence of differential gene expression during larval development (**20**). In contrast dimorphic female naked mole-rats are produced by divergent growth involving reproductively mature individuals.

Besides species-specific behavioral variants connected with ecological niches and cultural strategies, the above string of data collectively provides a comparative, context-based evolutive perspective in which convergence of basic behavioral drives—survival, dominance/prevalence, and the prevalence of kin cooperation—takes place on a broad spectrum of invertebrate and vertebrate species. This includes the concept of hierarchical dominance strategies in social structures; the impact of social position in access to nutrients, reproductive rights, and labor classes; and their correlative impact on brain organization and development. These evolutive vectors appear masked in higher-order vertebrates, especially under complex cultural structures such as in human societies. That may be the case of the *right of pernada* during the Middle Ages (in medieval vulgar Latin, *ius primae noctis*, or *right of the first night*, or *droit du seigneur*). It refers to a presumed right that granted to the feudal lords the power to have sexual relations with any maid who was going to marry one of his servants. This *droit du seigneur* would find evolutive grounds based on a social structure in other mammalian species in which reproductive rights within a community were based on hierarchical grounds, though framed under a cultural context (see also in Colombo 2019).

References

Alward BA, Laud VA, Skalnik CJ, York RA, Juntti SA, Fernald RD. Modular genetic control of social status in a cichlid fish. PNAS. 2020;117:28167–74. https://doi.org/10.1073/pnas.2008925117.

Amrein I, Becker AS, Engler S, Huang S-h, Müller J, Slomianka L, Oosthuizen MK. Adult neurogenesis and its anatomical context in the hippocampus of three mole-rat's species. Front Neuroanat. 2014;8:1–11. https://doi.org/10.3389/fnana.2014.00039.

Barchuk AR, Cristino AS, Kucharski R, Costa LF, Simões ZLP, Maleszka R. Molecular determinants of caste differentiation in the highly eusocial honeybee *Apis mellifera*. BMC Dev Biol. 2007;7:70. https://doi.org/10.1186/1471-213X-7-70.

Barker AJ, Veviurko G, Bennett NC, Hart DW, Mograby L, Lewin GR. Culturaltransmission of vocal dialect in the naked mole-rat. Science. 2021;371:503–7. https://doi.org/10.1126/science.abc6588.

Berreman GD. Caste as social process. Southwest J Anthropol. 1967;23:4.

Bourke AFG. The validity and value of inclusive fitness theory. Proc R Soc B. 2011;278:3313–20. https://doi.org/10.1098/rspb.2011.1465.

Brand N, Chapuisat M. Born to be bee, fed to be worker? The caste system of a primitively eusocial insect. Front Zool. 2012;9:35. http://www.frontiersinzoology.com/content/9/1/35

Burkart JM, Schubiger MN, van Schaik CP. The evolution of general intelligence. Behav Brain Sci. 2017;40:e195. https://doi.org/10.1017/S0140525X16000959.

Canciani M, Arnellos A, Moreno A. Revising the superorganism: an organizational approach to complex eusociality. Front Psychol. 2019; https://doi.org/10.3389/fpsyg.2019.02653.

Chapman TB, Crespi BJ, Kranz BD, Schwarz MP. High relatedness and inbreeding at the origin of eusociality in gall-inducing thrips. Proc Natl Acad Sci U S A. 2000;97(4):1648–50. https://doi.org/10.1073/pnas.020510097.

Colombo JA. Pobreza y desarrollo infantil (Poverty and child development). Buenos Aires: Ed. Paidos; 2007.

Colombo JA. Our animal condition and social construction. New York: Nova Science Publishers; 2019. ISBN: 978-1-53615-357-6. eBook ISBN: 978-1-53615-3583

Crespi BJ, Yanega D. The definition of eusociality. Behav Ecol. 1995;6:109–15. https://doi.org/10.1093/beheco/6.1.109.

Foster KR., Ratnieks LW. A new eusocial vertebrate? Trends Ecol. Evol. 20: 363–364, 2005.

Gheusi G, Ortega-Perez I, Murray K, Lledo P-M. A niche for adult neurogenesis in social behavior. Behav Brain Res. 2009;200:315–22.

Goodson JL. The vertebrate social behavior network: evolutionary themes and variations. Horm Behav. 2005;48(1):11–22. https://doi.org/10.1016/j.yhbeh.2005.02.003.

Hardisty BE, Cassill DL. Extending eusociality to include vertebrate family units. Biol Philos. 2010;25:437–40. https://doi.org/10.1007/s10539-009-9176-8.

Holmes MM, Rosen GJ, Jordan CL, de Vries GJ, Goldman BD, Forger NG. Social control of brain morphology in a eusocial mammal. PNAS. 2007;104:10548–52. https://doi.org/10.1073/pnas.0610344104.

Jarosch A, Stolle E, Crewe RM, Moritz RFA. Alternative splicing of a single transcription factor drives selfish reproductive behavior in honeybee workers (*Apis mellifera*). PNAS. 2011;108(37):15282–7. https://doi.org/10.1073/pnas.110934.

Jarvis JUM, Sherman PW. Heterocephalus Glaber. Mamm Species. 2002;706:1–9.

Kapheim KM, Nonacs P, Smit AR, Wayne RK, Wcislo WT. Kinship, parental manipulation and evolutionary origins of eusociality. Proc R Soc B. 2015;282:20142886. https://doi.org/10.1098/rspb.2014.2886.

Kozorovitskiy Y, Gould E. Dominance hierarchy influences adult neurogenesis in the dentate gyrus. J Neurosci. 2004;24:6755–9.

Kverkova K, Bělkov T, Olkowicz S, Pavelkov Z, O'Riain MJ, Šumbera R, Burda H, Bennett NC, Němec P. Sociality does not drive the evolution of large brains in eusocial African mole-rats. Sci Rep. 2018;8:9203. https://doi.org/10.1038/s41598-01826062-8.

Lacey EA, Sherman PW. Redefining eusociality: concepts, goals, and levels of analysis. Ann Zool Fenn. 2005;42:573–7.

Lahdenpera M, Lummaa V, Helle S, Tremblay M, Russell AF. Fitness benefits of prolonged post-reproductive lifespan in women. Nature. 2004;428:178–81.

Lattorff HMG, Moritz RFA, Crewe RM, Solignac M. Control of reproductive dominance by the lytoky gene in honeybees. Biol Lett. 2007;3:292–5. https://doi.org/10.1098/rsbl.2007.0083.

Lipina SJ, Colombo JA. Poverty and brain development during childhood. Washington, DC: American Psychological Association; 2009.

Michener CD, Brothers DJ. Were workers of eusocial hymenoptera initially altruistic or oppressed? Proc Natl Acad Sci. 1974;71:671–4. https://doi.org/10.1073/pnas.71.3.671.

Moda LM, Vieira J, Guimarães Freire AC, Bonatti V, Bomtorin AV, Barchuk AG, Simões ZLP. Nutritionally driven differential gene expression leads to heterochronic brain development in honeybee castes. PLoS One. 2013;8(5):e64815. https://doi.org/10.1371/journal.pone.0064815.

Moritz RFA, Lattorff HMG, Crous KL, Hepburn RH. Social parasitism of queens and workers in the Cape honeybee (Apis mellifera capensis). Behav Ecol Sociobiol. 2011;65:735–40. https://doi.org/10.1007/s00265-010-1077-y.

Newman SH. The medial extended amygdala in male reproductive behavior a node in the mammalian social behavior network. Ann N Y Acad Sci. 1999;877:242–57.

Nowak MA, Tarnita CE, Wilson EO. The evolution of eusociality. Nature. 2010;466:1057–62. https://doi.org/10.1038/nature09205.

O'Riain MJ, Jarvis JUM, Alexander R, Buffenstein R, Peeters C. Morphological castes in a vertebrate. PNAS. 2000;97:13194–7. https://doi.org/10.1073/pnas.97.24.13194.

Oya S, Kohno H, Kainoh Y, Ono M, Kubo T. Increased complexity of mushroom body Kenyon cell subtypes in the brain is associated with behavioral evolution in hymenopteran insects. Sci Rep. 2017;7:13785. https://doi.org/10.1038/s41598-017-14174-6.

Pahlke S, Jaumann S, Seid MA, Smith AR. Brain differences between social castes precede group formation in a primitively eusocial bee. Sci Nature. 2019;106:49. https://doi.org/10.1007/s00114-019-1644-7.

Perez-Barbera FJ, Shultz S, Dunbar RIM. Evidence for coevolution of sociality and relative brain size in three orders of mammals. Evolution. 2017;61(12):2811–21. https://doi.org/10.1111/j.1558-5646.2007.00229.

Powell S, Franks NR. Caste evolution and ecology: a special worker for novel prey. Proc R Soc B. 2005;272:2173–80. https://doi.org/10.1098/rspb.2005.3196.

Ratnieks FLW, Helantera H. The evolution of extreme altruism and inequality in insect societies. Philos Trans R Soc B. 2009;364:3169–79. https://doi.org/10.1098/rstb.2009.0129.

Scantlebury M, Speakman JR, Oosthuizen MK, Roper TJ, Bennett NC. Energetics reveals physiologically distinct castes in a eusocial mammal. Nature. 2006;440:795–7. https://doi.org/10.1038/nature04578.

Shell WA, Rehan SM. Social modularity: conserved genes and regulatory elements underlie caste antecedent behavioural states in an incipiently social bee. Proc R Soc B. 2019;286:20191815. https://doi.org/10.1098/rspb.2019.1815.

Shell WA, Steffen MA, Pare HK, Seetharam AS, Severin AJ, Toth AL, Rehan SM. Sociality sculpts similar patterns of molecular evolution in two independently evolved lineages of eusocial bees. Commun Biol. 2021;4:253. https://doi.org/10.1038/s42003-021-01770-6.

Sherman PW, Lacey EA, Reeve HK, Keller L. The eusociality continuum. Behav Ecol. 1994;6:102–8.

Tian L, Zhou X. The soldiers in societies: defense, regulation, and evolution. Int J Biol Sci. 2014;10:296–308. https://doi.org/10.7150/ijbs.6847.

Vaiserman A. Developmental epigenetic programming of caste specific differences in social insects: an impact on longevity. Curr Aging Sci. 2014;7(3):176–86. https://doi.org/10.2174/1874609807666141129173749.

Warner MR, Qiu L, Holmes MJ, Mikheyev AS, Linksvayer TA. Convergent eusocial evolution is based on a shared reproductive ground plan plus lineage-specific plastic genes. Nat Commun. 2019;10(2651):1–11.

Weiner SA, Toth AL. Epigenetics in social insects: a new direction for understanding the evolution of castes. Genet Res Int. 2012;Article ID 609810:1–11. https://doi.org/10.1155/2012/609810.

Wilson EO, Holldobler B. Eusociality: origin and consequences. PNAS. 2005;102(13367):13371.

Wojciechowski M, Lowe R, Maleszka J, Conn D, Maleszka R, Hurd PJ. Phenotypically distinct female castes in honeybees are defined by alternative chromatin states during larval development. Genome Res. 2018;28(10):1532–42. https://doi.org/10.1101/gr.236497.118.

Chapter 3
Brain and Social Environment

> The evolution of castes in a mammal and insects represents a striking example of convergent evolution for enhanced fecundity in societies characterized by high reproductive skew. Similarities in the selective environment between naked mole-rats and eusocial insect species highlight the selective conditions under which queen/worker castes are predicted to evolve in animal societies. (O'Riain et al. 2000)

> The hallmark of eusociality is reproductive division of labour between generations, a surprising social organization by which some individuals become functionally sterile helpers [1]. (cf., Wilson, 1967; quoted by Brand and Chapuisat 2012)

> Although behavioral neuroendocrinologists often discuss "sociality" as a unitary variable, the term encompasses a wide diversity of behaviors that do not evolve in a linked fashion across species. Thus grouping, monogamy, paternal care, cooperative breeding/alloparental care, and various other forms of social contact are evolutionarily labile and evolve in an almost cafeteria-like fashion, indicating that relevant neural mechanisms are at least partially dissociable. (Goodson 2013)

Brain Correlates of Social Standings Under *Dominance* and Enriched Social Conditions

Brain neural activity and connectivity and mental activity constantly interact with environmental, emotional, and introspective domains. In this multivariate condition, sociality—relative positioning within hierarchical community structures, environmental and relational inputs, and access to feeding resources—plays a significant role in brain and mental development and, to a significant extent, on individual cognitive and emotional profiles expressed in adult life. Hence, in considering comparative approaches, it is necessary to consider historical adaptation to the natural ecological niche, which incorporates a concept of relativity in terms of evaluating cognitive resources, as stated by Salwiczek et al. (2012):

> …animals' cognitive abilities are linked to their evolutionary history, and hence their ecology…

J. A. Colombo, *Dominance Behavior*, https://doi.org/10.1007/978-3-030-97401-5_3

The above statements are supported by human and nonhuman observational and experimental studies (Colombo 2019). The previous chapter described the impact of early (larval) feeding in defining castes—and brain correlates—in eusocial species.

Hence, regardless of the considered species, material and emotional conditions during early development and insertion within a given social structure affect individual degrees of freedom and brain/mental potential during adult life. Due to comparative relative neural plasticity across species, under given circumstances—mostly in human domains due to its prolonged postnatal period of brain maturation—this could be modified whenever means are implemented in due time and form. The following paragraphs provide evidence for these assertions whatever the social structure might be—considering the relative permissiveness of different sociocultural structures among the different species, of human societies in particular. Within variable degrees among animal species, they underscore the universality of early raising conditions in molding or defining the probability of individual profile and social insertion. The spread and expression of the menu of social and behavioral variables involved will depend on the species' genome considered and on individual and collective social standings and resilience.

In the year 2012, Zayed and Robinson, based on research performed on the honeybee, reported that:

> ...changes in brain gene expression mediate changes in behavior, and that the association between specific genes and behavior exists over multiple timescales, from physiological to evolutionary.

Additionally, support on behavioral changes based on insect social learning was provided by Leadbeater and Dawson (2017) when they state that:

> An expanding body of work now shows that social learning (1), once considered the preserve of vertebrate species, is a feature of insect behavioral repertoires (2, 3). Insects not only learn about foraging skills, food preferences, brood hosts, and potential mates by responding to information provided inadvertently by others (4–11), but also transmit these behaviors further, such that they propagate through groups (8, 12) and possibly even through wild populations (13). Some of these phenomena appear similar to socially learned behavior patterns that have been described in vertebrates (11, 14) at least outside the context of imitation, and as such they are interesting extensions to the taxonomic distribution of social learning.

As mentioned in Chap. 2, according to Moda et al. (2013), following studies performed in *Apis mellifera* during the larval period, the brains of queens grow larger and faster, representing developmental heterochrony due to a differential feeding regime. It involves caste-specific transcriptional profiles of a set of genes, which mediates differential nutrition and neurogenesis, as stated by Moda et al. (2013):

> Thus, pointing to a link between differential nutrition and differential neurogenesis via genes that control cell proliferation and fasciculation.

Similar statements by other authors, as shown below, reinforce the concept that brain caste differentiation implies early feeding strategies. These observations underline the role of sociality on the probabilities of establishing differential brain developmental conditions through actions applied at early developmental stages, which seems to define a universal condition (see later in humans).

In naked mole rats, Kverkova et al. (2018) draw careful conclusions on their observations, as they first state that:

> These subterranean rodents exhibit a broad spectrum of social complexity, ranging from strictly solitary through to eusocial cooperative breeders, but feature similar ecologies and life history traits. We found no positive association between sociality and neuroanatomical correlates of information-processing capacity. Solitary species are larger, tend to have greater absolute brain size and have more neurons in the forebrain than social species. The neocortex ratio and neuronal counts correlate negatively with social group size… These findings suggest that group living or pair bonding per se does not select strongly for brain enlargement unless coupled with Machiavellian interactions affecting individual fitness.

This is followed by a cautious ending statement:

> …our findings cannot rule out that sociality does select for larger brains in mole-rats, as all we can observe is the end result of all selective pressures and constraints put together.

According to studies performed by Amrein et al. (2014), in this same species, the percentages of new neurons are high in the Highveld and naked mole rats and lower in solitary *Cape mole* rats, thus supporting that neuronal differentiation reflects species-specific demands, which appear lower in subterranean rodents.

Holmes et al. (2007) draw a general conclusion on the role of social factors on the development of naked mole rats. According to these authors, naked mole rats (*Heterocephalus glaber*) exhibit the strictest reproductive hierarchy known to mammals and the closest mammalian equivalent of eusociality. Authors conclude that status rather than sex predominates in determining neural structure.

These, and former quoted observations, lead to considering the role of early social and nutritional factors acting upon individual development on mammalian species. Subjacent to this variability in behavioral profiles depending on early rising conditions persists the ancestral survival drive fueled by dominance and gain (profit)-seeking behaviors.

Enriched Environment and Development

Regarding dominance hierarchies in naked mole rats, Jarvis and Sherman (2002) state that:

> Size-based dominance hierarchies occur within colonies (Clarke and Faulkes, 1998; Schieffelin and Sherman, 1995). The largest individuals of both sexes, especially the breeders, obtain access to food or digging sites before the smaller colony mates.

More generally, among mammalian species, since the pioneering work of Altman and Das (1964), enriched environments are known to modify brain neuronal events, as shown in the visual cortex of rats by Jones and Greenough (1996). They observed that rats raised from weaning in a complex environment had an increased number of synapses per neuron in the visual cortex compared to animals housed in standard laboratory cages. Those animals also showed increased synaptic contact of astroglia, suggesting an experience-related enhancement of the astrocytic involvement in synaptic activity.

According to Uylings et al. (1978), environmental influences affect the cerebral cortex structure of rats, as shown by an increase in branching and the length of dendritic processes in pyramidal neurons in the visual cortex of adult rats.

Kempermann et al. (1997) observed that an enriched environment results in significantly more new neurons in the dentate gyrus of the hippocampus in mice, as compared with littermates housed in standard cages, which is associated with improved spatial memory. Hence, the mentioned neural structural changes result in functional ones, such as spatial memory, as Nilsson et al. (1999) also mentioned.

Later, Edwards and Spitzer (2006) analyzed in crayfish reversible environmental effects on establishing social behavior-dominance hierarchy. The authors described conceptually interesting effects of hierarchical standings, depending on their dominant or subordinate role:

> In crayfish, social interactions between pairs of animals lead to changes in behavior that mark the formation of a dominance hierarchy. Those changes in behavior result from changes in the excitability of specific neural circuits. In the new subordinate, circuits for offensive behavior become less excitable and those for defensive behavior become more excitable. Serotonin, which is implicated in mechanisms for social dominance in many animals, modulates circuits for escape and avoidance responses in crayfish. The modulatory effects of serotonin on the escape circuits have been found to change with social dominance, becoming excitatory in dominant crayfish and inhibitory in subordinates.
>
> …these changes occur over a 2-week period and for the subordinate are reversible at any time following a reversal of the animal's status.

More recently, also in crayfish, Issa et al. (2012) reinforced the above observations on the interaction of social status and serotoninergic modulatory components, according to the following statement:

> These results indicate that the circuits and their intrinsic serotonergic modulatory components are configured differently according to social status…

This set of observations addresses one of the main topics of this essay, the comparative emergence of dominance or prevalence in social structures, from insects to mammals, and its impact on individual and collective phenotypes and on brain organization.

Considering that hierarchical status and dependence generate different conditions for social interaction, stress exposure and management, and nutritional access, it is pertinent to note that reports in rodents underline the significance of social rank on social stress. Kozorovitskiy and Gould (2004), following their research in rats, stated that:

> … more new neurons were observed in the dentate gyrus of the dominant males compared with both subordinates and controls… social status rather than living in a complex environment account for the effect of dominance on adult neurogenesis.
>
> …we show that social dominance does not alter cell proliferation but is associated with enhanced survival of new neurons in the dentate gyrus.

Beery et al. (2020) underlined the role of social ranks and stress intensity:

> Social stress is not uniform across all individuals in a social group across all time periods. In both rats and mice, social rank significantly influences the timing and intensity of social stress experienced by individuals. Subordinate animals typically show elevated stress responsivity and long-term negative effects of stress on their behavior and immune, cardiovascular, metabolic and reproductive functioning.

This brief account of observations on the impact of social structure on individual neurobehavioral construct among various species underlines that although species-specific social structure and response to social hierarchies vary, the latter shows a widespread impact on neurobiological components involved in individual behavioral adaptation and performance. In more complex societies, in which social strata subsist—whether economic, religious, or labor-bound based—cultural contexts and instruments mediate these interactions. Apropos of this statement, the following excerpt regarding the social behavior network seems appropriate to be mentioned:

> Based on a variety of behavioral, connectional and histochemical findings, it is now clear that the brain circuits that regulate social behavior in non-mammalian vertebrates are extensively similar to those in mammals. These observations strongly support the proposal that **the social behavior network** (as first described for mammals, Newman, 1999) **is a fundamental and evolutionarily conserved feature of the vertebrate brain.** (Goodson 2005) (Bold letters inserted by JAC)

Besides insects and nonmammalian vertebrate species (in fishes, see Alward et al. 2020), relative social standing, enriched environment, poverty conditions, and social hierarchies are all variables that could be considered as universals, based on the known impact on brain development and cognitive performance in mammalian species, including humans, as summarized in Colombo (2019):

> Several experimental studies followed those initially performed by Harlow (1959) and the groups of Greenough and Rosenzweig (Krech et al. 1960; Volkmar and Greenough, 1972; Turner and Greenough, 1985). They confirm that the physical and social conditions of the environment during the breeding period and later affect brain development. This takes place at the microstructural and neurochemical level; a condition that influences cognitive abilities and emotional behavior in adult life.

As it will be further expanded later in humans, it may be worth considering the impact of sociocultural and early experience context on brain configuration and cognitive and behavioral patterns affecting the dynamic of social structures. Lieberwirth and Wang have also mentioned this (2012):

> … it seems that acute and chronic sociosexual interactions, as positive stimuli, facilitate cell proliferation and survival across distinct brain regions; whereas aversive social interactions leading to psychosocial stress impair adult neurogenesis.

The concept of the impact of environmental effects on individual development, what in modern terms is called epigenetics, was advanced by C. Darwin (1874), as expressed in the following quotation:

> … my critics frequently assume that I attribute all changes of corporeal structure and mental power exclusively to the natural selection pf such variations as are often called spontane-

ous; whereas, even in the first edition of the "Origin of species", I distinctly stated that great weight must be attributed to the inherited effects of use and disuse, with respect both to the body and mind. I also attributed some amount of modification to the direct and prolonged action of changed conditions of life.

References

Altman J, Das GD. Autoradiographic examination of the effects of enriched environment on the rate of glial multiplication in the adult rat brain. Nature. 1964;204:1161–3.

Alward BA, Laud VA, Skalnik CJ, York RA, Juntti SA, Fernald RD. Modular genetic control of social status in a cichlid fish. PNAS. 2020;117:28167–74. https://doi.org/10.1073/pnas.2008925117.

Amrein I, Becker AS, Engler S, Huang S-h, Müller J, Slomianka L, Oosthuizen MK. Adult neurogenesis and its anatomical context in the hippocampus of three mole-rat's species. Front Neuroanat. 2014;8:1–11. https://doi.org/10.3389/fnana.2014.00039.

Beery AK, Holmes MM, Lee W, Curley JP. Stress in groups: lessons from non-traditional rodent species and housing models. Neurosci Biobehav Rev. 2020;113:354–72. https://doi.org/10.1016/j.neubiorev.2020.03.033.

Brand N, Chapuisat M. Born to be bee, fed to be worker? The caste system of a primitively eusocial insect. Front Zool. 2012;9:35. http://www.frontiersinzoology.com/content/9/1/35

Colombo JA. Our animal condition and social construction. New York: Nova Science Publishers; 2019. ISBN: 978-1-53615-357-6. eBook ISBN: 978-1-53615-3583

Darwin C. The descent of man and selection in relation to sex. Preface to the second edition. 2nd ed. London; 1874.

Edwards DH, Spitzer N. Social dominance and serotonin receptor genes in crayfish. Curr Top Dev Biol. 2006;74:177–99. https://doi.org/10.1016/S0070-2153(06)74006-6.

Goodson JL. The vertebrate social behavior network: evolutionary themes and variations. Horm Behav. 2005;48(1):11–22. https://doi.org/10.1016/j.yhbeh.2005.02.003.

Goodson JL. Deconstructing sociality, social evolution, and relevant nonapeptide functions. Psychoneuroendocrinology. 2013;38(4):465–78. https://doi.org/10.1016/j.psyneuen.2012.12.005.

Holmes MM, Rosen GJ, Jordan CL, de Vries GJ, Goldman BD, Forger NG. Social control of brain morphology in a eusocial mammal. PNAS. 2007;104:10548–52. https://doi.org/10.1073/pnas.0610344104.

Issa FA, Drummond J, Cattaert D, Edwards DH. Neural circuit reconfiguration by social status. J Neurosci. 2012;32:5638–45.

Jarvis JUM, Sherman PW. Heterocephalus Glaber. Mamm Species. 2002;706:1–9.

Jones TA, Greenough W. Ultrastructural evidence for increased contact between astrocytes and synapses in rats reared in a complex environment. Neurobiol Learn Mem. 1996;65:48–56.

Kempermann G, Kuhn HG, Gage FH. More hippocampal neurons in adult mice living in an enriched environment. Nature. 1997;386:493–5.

Kozorovitskiy Y, Gould E. Dominance hierarchy influences adult neurogenesis in the dentate gyrus. J Neurosci. 2004;24:6755–9.

Kverkova K, Bělkov T, Olkowicz S, Pavelkov Z, O'Riain MJ, Šumbera R, Burda H, Bennett NC, Němec P. Sociality does not drive the evolution of large brains in eusocial African mole-rats. Sci Rep. 2018;8:9203. https://doi.org/10.1038/s41598-01826062-8.

Leadbeater E, Dawson EH. A social insect perspective on the evolution of social learning mechanisms. PNAS. 2017;114:7838–45.

Lieberwirth C, Wang Z. The social environment and neurogenesis in the adult mammalian brain. Front Hum Neurosci. 2012;6:1–19. https://doi.org/10.3389/fnhum.2012.00118.

Moda LM, Vieira J, Guimarães Freire AC, Bonatti V, Bomtorin AV, Barchuk AG, Simões ZLP. Nutritionally driven differential gene expression leads to heterochronic brain development in honeybee castes. PLoS One. 2013;8(5):e64815. https://doi.org/10.1371/journal.pone.0064815.

Nilsson M, Perlieva E, Johansson U, Orwar O, Eriksson PS. Enriched environment increases neurogenesis in the adult rat dentate gyrus and improves spatial memory. J Neurobiol. 1999;39:569–78.

O'Riain MJ, Jarvis JUM, Alexander R, Buffenstein R, Peeters C. Morphological castes in a vertebrate. PNAS. 2000;97:13194–7. https://doi.org/10.1073/pnas.97.24.13194.

Salwiczek LH, Pretot L, Demarta L, Proctor D, et al. Adult cleaner wrasse outperform capuchin monkeys, chimpanzees, and orangutans in a complex foraging task derived from cleaner—Client Reef Fish Cooperation. PLoS One. 2012;7:1–9, e49068. https://doi.org/10.1371/journal.pone.0049068.

Uylings HBM, Kuypers K, Veltman WAM. Environmental Influences on the Neocortex in Later Life. Prog Brain Res. 1978;48:261–74. https://doi.org/10.1016/S0079-6123(08)61028-6.

Zayed A, Robinson GE. Understanding the relationship between brain gene expression and social behavior: lessons from the honeybee. Annu Rev Genet. 2012;46:591–615. https://doi.org/10.1146/annurev-genet-110711-15551.

Chapter 4
Sociobiological Interactions

> Alternatively, definitional problems could be reduced by expanding the eusociality concept to include all vertebrate and invertebrate societies with helpers. We favor the latter approach. It seems more productive to recognize that similar social systems occur in birds, mammals, and insects than to debate whether particular insects are eusocial (e.g., Furey, 1992 versus Tsuji, 1992). (Sherman et al. 1995)

> Social hierarchies are ubiquitous in social species and profoundly influence physiology and behavior. (Alward et al. 2020)

Considering evidence gathered from nonhuman studies on social and physical environmental interactions acting on behavioral and developmental events, the question emerges as to whether common elements arise in human neurobiology and behavior.

As mentioned previously, the following brief statements suggest a universal biological constant in terms of social/environmental impact and social inequalities on individual neural and behavioral construction, one primary subject of this essay.

> …the heritable division of labor within species, involving different types or morphs, is relatively rare except among eusocial insects. This rarity within species is explained by the Bishop-Cannings theorem, which shows that any division of labor that leads to mean fitness differences among occupational types will be unstable because the type with higher fitness should outcompete the type with lower fitness (Bishop and Cannings 1978). Different types may persist at equilibrium, but they must have the same mean fitness (payoff).
>
> …larger, more densely connected populations are likely to produce faster cultural evolutionary rates for sophisticated technology, complex skills, and knowledge (Henrich 2004b; Shennan 2001). This, in turn, generates more surplus (Carneiro and Tobias 1963) and favors stratification, permitting greater degrees of societal inequality to emerge. (Henrich and Boyd 2008)

> …social animals can be defined as those who cannot survive alone and rely on members from their group to regulate their ongoing physiology (or allostasis) (*i.e., the process of maintaining homeostasis through the adaptive change of the organism's internal environment to meet perceived and anticipated demands* [1]). The rather simple evolutionary constraint of social dependency for survival can be sufficient to make the social environment

[1] https://www.sciencedirect.com/topics/immunology-and-microbiology/allostasis

J. A. Colombo, *Dominance Behavior*, https://doi.org/10.1007/978-3-030-97401-5_4

vitally salient, and to provide the ultimate driving force for socially crafted brain development and learning. (Atzil et al. 2018) (Italics and footnote added by JAC)

Social interactions may shape brain development. In primitively eusocial insects, the mushroom body (MB), an area of the brain associated with sensory integration and learning, is larger in queens than in workers. This may reflect a strategy of neural investment in queens or it may be a plastic response to social interactions in the nest... Larger MBs among foundresses may reflect the increased larval nutrition provisioned to future queens and the lack of social aggression from a dominant queen upon adult emergence. (Pahlke et al. 2019)

The above statements concerning socioecological influences on neurobiological domains can thus be further analyzed, as proposed below.

Sociobiological Interactions

Brain neurogenesis and neural circuit ensemble formation undergo developmental stages that depend on brain complexity and socioecological (habitat) conditions of the considered species and brain regions. During this species-specific developmental timing, the brain traverses through varying stages of susceptibility to be affected by environmental factors of either social or physical nature. These epigenetic[2] domains (Colombo 2019) impact brain configuration, its performance, and behavioral profiles, as described in insects and mammalian species (in mammalian species see Harlow 1959; Krech et al. 1960; Altman and Das 1964; Volkmar and Greenough 1972; Uylings et al. 1978; Turner and Greenough 1985; and Hackman and Farah 2009, Hackman et al. 2010; in insects see Barchuk et al. 2007; Moda et al. 2013; Lieberwirth and Wang 2012; Weiner and Toth 2012; and Alward et al. 2020).

As stated by Fang and Yuste (2017) regarding neuronal production level:

Our work thus demonstrates that neuronal production level can shape the functional modularity of the neocortex and regulate perceptual discrimination of animals.

... manipulation that increases neuronal number also increases functional modularity in the neocortex and perceptual discrimination.

Research in insects provides further insight into the impact of early developmental conditions, for example, regarding epigenetic factors involved in DNA expression. The work by Weiner and Toth (2012) coincides that:

...epigenetic modifications of DNA are important for mediating the effect of the early social (maternal) environment on adult phenotype (reviewed in [7, 20]). This work led to the suggestion that social modulation of the genome, and the resulting adult plasticity, may

[2] Epigenetics. "...we refer specifically to chemical modifications to DNA that do not change the DNA sequence." "...modifications can be made to the histones around which DNA is packaged [20]. These modifications include methylation, acetylation, and ubiquitination [32]. All these different modifications have the potential to affect transcription via changes in chromatin structure and/or gene-splicing patterns [20, 23, 32]." (Weiner and Toth 2012)

> rely heavily on epigenetic effects [20] … In our conceptual framework (Figure 4.1), we propose that dietary differences lead to differential methylation. This, in turn, leads to alternative splicing and possibly caste-biased expression, which leads to caste biased phenotypes, such as restricted ovarian development in workers or larger body size and longer lifespan in queens.

In mammals, on a neurobiological basis, as explored in mice by Yamamuro et al. (2020), sociability would be interactively regulated by a circuit composed of prefrontal projecting pathways to the posterior paraventricular thalamus, presumably involving somatostatin.

If we consider that in modern times the concept of an ecological domain should include the sociocultural environment—besides the natural physical domain probably predominant in early times—the question arises as to the current drives on neural dynamics that impinge on the neural plasticity of our brains. Interactions (dominant or prevalent vs. submissive) among individuals of animal social groups represent an intrinsic component of the "environmental" or ecological context. As such, dissecting these components from the ecological context may not capture the complex reality of animal behavioral evolution. In this regard, the physical component of ecology in most contemporary human societies has significantly given way to sociocultural ecology. The question arises as to its impact and relative predominance on brain organization and its continuous evolution, specifically, for our concern, regarding long-term social inequities imposed on several human populations.

Basic behavioral patterns are acquired at early ages in vertebrates, while social interactions provide key wiring instructions that contribute to determining brain development (Atzil et al. 2018). In this regard, neurogenesis in adults represents a constitutive replacement mechanism for lost neurons, as well as a process involved in neural plasticity in response to specific experiences throughout life, as described by Gheusi et al. (2009):

> Although chronic subordination is associated with a lower rate of cell production within the adult dentate gyrus (*hippocampus*), social interactions can also provide protective effects against negative consequences of experience-induced glucocorticoids on adult neurogenesis. First, social isolation decreases hippocampal cell proliferation in individually housed rats. This isolation-induced decrease in cell proliferation can be reversed by subsequently housing the animals in groups [69], thus indicating that social housing represents a key regulator of hippocampal neurogenesis. (Italics inserted by JAC)

Genes and Social Interactions

The widespread impact of culture on genetic evolution in nonhuman species was pointed out by Whitehead et al. (2019):

> Early reviews assumed that non-human cultures were insufficiently stable to affect genetic evolution[1,2]. However, recent research has established that animal culture is present in insects, birds, fishes and mammals[3,4], that it can have important impacts on fitness (e.g., ref.[5]), that it can be stable over many generations (e.g., refs.[6,7]), and that it can affect evolutionary dynamics even when transient (e.g., refs.[8,9]).

Jones et al. (2020) explored the plasticity in brain gene regulatory networks in honeybees as a function of their social role. These authors established that foragers and egg layers exhibited widespread differences in brain gene expression, with differential expression of nearly half (46%) of all genes expressed in the brain. Furthermore, according to these authors, continuous phenotypic variation was found associated with continuous variation in brain gene expression, brain chromatin accessibility, and transcription factor activity, the latter providing individual predictive behavioral phenotype. This led to the conclusion that incremental changes in gene regulatory network activity led to decoupling solitary behavioral programs into a distinct queen and worker castes. This hypothesis could provide a framework for understanding the evolution of eusociality at the molecular level.

Maruska et al. (2012) used an African cichlid fish[3] to test the hypothesis that a novel social opportunity activates the conserved "social behavior network," a collection of brain nuclei that regulates social behaviors among vertebrates. The authors measured mRNA levels of immediate early genes[4] in brain regions as a proxy for neuronal activation. Levels were higher in all nuclei of the "social behavior network" in males that were allowed to rise in social rank compared to stable subordinate and dominant individuals.

According to Miller et al. (2017), differences in behavior patterns of dominant and subordinate animals of many species suggest widespread social regulation of neural circuits. Additionally, applying experimental and computational modeling, the authors report that:

> Using zebrafish as a model system, we demonstrate how social experience affects decision making to enable animals to produce socially appropriate behavior. Based on experimental evidence and computational modeling, we show that behavioral decisions reflect the interplay between competing neural circuits whose activation thresholds shift in accordance with social status.

The impact of sociobiological interactions on genes controlling mammalian encephalization and the perspectives of individual development across several orders and species, as summarized here, recalls the need to review these concepts within the domain of human sociobiological interactions as it will be considered later. In general, as proposed by Schultz and Dunbar (2010) in mammals:

> …high encephalization is associated specifically with sociality (and particularly stable groups). This suggests that the pressure for increased encephalization is associated with some aspect of bonded sociality.

[3] Freshwater fish. Generally, cichlids tend to be of medium size, ovate in shape, and slightly laterally compressed.

[4] Genes which are activated transiently and rapidly in response to a wide variety of cellular stimuli. They represent a standing response mechanism that is activated at the transcription level in the first round of response to stimuli, before any new proteins are synthesized.

In terms of the functional organization, as it will be considered later, cognitive development depends on emotional and physical contexts (environmental enrichment) during early postnatal stages.

In insects, according to Michener and Brothers (1974), the evidence indicates that the primary factor leading to the appearance of a worker caste resides in control by the queen based on feeding rights (altruism induced by "oppression"). Individuals foregoing reproduction assist others in taking place in communal breeding species. According to Scantlebury et al. (2006), this phenomenon is seen at its most extreme in eusocial organisms, where reproductive "queens" are responsible for producing all the offspring of a colony and would represent the most extreme form of altruism known among nonhuman animals. According to O'Riain et al. (2000):

> Similarities in the selective environment between naked mole-rats and eusocial insect species highlight the selective conditions under which queen/worker castes are predicted to evolve in animal societies.

At the molecular, genetic level, Yamazakia et al. (2006) analyzed insect ecdysteroid-regulated genes selectively expressed in the honeybee brain's mushroom bodies (MBs). As previously mentioned, the MBs are insect brain regions essential for memory formation and sensory integration. The authors found a statistically significant difference in the relative mRNA levels between the foragers (increased levels) and nurse bees or queens.

Social Stratification and Altruism

As implied in the *evolutive Machiavelli imposition* of social stratification, privileged developmental conditions determine individual social ranks and roles. This could be projected as a secondary effect to ongoing inequalities in more complex social structures, such as those generated in human societies, a theme that merits further consideration, to be discussed later.

This form of imposed altruistic behavior on lower ranks occurs under queen dominance conditions where selective nutrient feeding represents a socially shared "privileged" medium, through which social hierarchies emerge. Fundamental to understand the behavioral organization, this privileged condition interacts with the genomic regulatory architecture of phenotypic plasticity, as it has been documented across a range of organisms and behavior, as stated by Zayed and Robinson (2012):

> … experiments demonstrate that brain gene expression is closely linked with behavior, that changes in brain gene expression mediate changes in behavior, and that the association between specific genes and behavior exists over multiple timescales, from physiological to evolutionary.

Regarding the concept of altruism in insects, it has been proposed by Ratnieks and Wenseleers (2007) a binary causal promotion, either as voluntary or enforced. At the human level, its supposedly voluntary dimension remains to be defined as to whether it is a subtle form of enforcement through values acquired during the individual social history that mold or promote this behavior. In this case, altruistic motivation would not surge spontaneously but either as an educational consequence of an anticipated product of the balance of gain and losses. It has been considered (Colombo 2019) that in humans' social values, altruism would not emerge from a *tabula rasa* but on experiential events among kin and foes.

According to Moda et al. (2013), in eusocial insects, the reported differential development is characterized by caste-specific transcriptional profiles of a set of genes. Thus, it points to a link between differential nutrition and differential neurogenesis via genes that control cell proliferation.

As stated by these authors:

> In conclusion, we showed that queen larvae develop larger brains than worker larvae and that the queen brain develops more rapidly than that of workers, thus representing a form of developmental heterochrony, reflecting the effect of the differential feeding regime of the two castes on nervous system development. We also showed that **this differential brain development is characterized by caste-specific transcriptional profiles for a set of genes** (APC-4, GlcAT-P, fax, kr-h1 and shot), **similar to the classically proposed explanation for the brain/body weight ratio favoring humans compared to chimpanzees** [50,51]. (Bold letters added by JAC)

According to Jones et al. (2020), using principal component analysis procedures, it was revealed that the degree of individual behavioral specialization was significantly correlated with measures of both brain gene expression and chromatin accessibility involving gene regulatory networks.

Stress and Social Environment

An additional dimension that emerges in mammalian stratified societies is when social conditions (social standing) influence individual stressful conditions. According to Beery et al. (2020), subordinate animals typically show elevated stress responsivity. This condition has long-term adverse effects on their behavior, immune, cardiovascular, metabolic, and reproductive functioning. As mentioned by the authors, group factors such as competition level, size of groups, group composition, space available, and the dominant individual's despotism affect its severity. These conditions are often combined with cultural evolutionary rates involving sophisticated technology, complex skills, and knowledge to favor stratification and social inequalities.

Besides its impact on social standings and probability of individual development, social inequalities often generate stressful conditions. These may affect neurobiological dynamics, e.g., hippocampal structure and function and inhibiting new cell birth in the dentate gyrus of the adult hippocampus, as has been shown in several species including primates (Gould et al. 1997, 1998; Eriksson et al. 1998; Nilsson et al. 1999; Kozorovitskiy and Gould 2004; Yap et al. 2006; Lieberwirth and Wang

2012). Interestingly, in three species of naked mole rats, subterranean living (i.e., with reduced environmental interactions) impacts hippocampal development, as revealed by reduced numbers of proliferating cells and young neurons. In these cases, neuronal differentiation reflects species-specific demands, which appear lower in subterranean rodents (Amrein et al. 2014). According to Holmes et al. (2007), social status (breeders as opposed to subordinates) rather than sex, drives neuronal number differences in the ventromedial nucleus of the hypothalamus and a larger volume of the bed nucleus of the stria terminalis, paraventricular nucleus, and medial amygdala.

In this respect, a general view is summarized by Issa et al. (2012):

> Social status affects neurogenesis in rodents (Kozorovitskiy and Gould, 2004) and crayfish (Song et al. 2007), neuronal size in fish (White et al. 2002), brain morphology in wasps (O'Donnell et al. 2007) and naked mole rats (2007), and cell receptor populations in crayfish (Spitzer et al. 2005) and fish (Burmeister et al. 2007). Social status also affects the serotonergic neuromodulation of synaptic responses in both crayfish (Yeh et al. 1996, 1997) and fish (Whitaker et al. 2011), and the excitability of neural circuits that produce different behaviors (Krasne et al. 1997; Herberholz et al. 2001; Neumeister et al. 2010.

According to Beery and Kaufer (2015), regarding physical and emotional stress effects, stress resilience varies with early life social environment—particularly with experience of maternal behavior and life history of exposure to mildly stressful experiences. At the human level, this acquires particular significance in terms of chronic stress of modern life and multiple stressors acting at all ages, with a particularly negative impact in poverty and marginal conditions.

Prosociality and Altruism: Direct Enforcement or Hedonistic Component? From Insects to Primates

> Altruistic traits, a term first coined by Haldane (1932), are those that decrease the fitness of their carrier but increase that of other individuals in the population.
>
> … Dawkins (1976) used the term *selfish gene* for any allele under natural selection, a re-emphasis of Hamilton's belief that altruistic traits could only evolve on the individual level, if they allow the spread of the selfish allele which determines them. (Eshel and Feldman 2001)

As with other domains of human culture, altruism raises issues regarding the origin of its human practice. The following considerations aim to propose possible evolutive paths for its emergence in our species, though cultural complexities of human behavior and socialization represent a particularly complex evolutionary event. Altruism is a behavior present in a broad spectrum of species, as it was described in insects in previous chapters. Across the various species, this behavior emerges as an imposed social profile, mediated by various species-specific dominance/prevalence strategies, following hierarchical power profiles (at the human level, caste, religious, emotional dependence, institutional). The concept of altruism is applied when a subject places a superior value in acting in favor of third parties or social status rather than on the risks he/she may endure when acting accordingly.[5] This behavior is observed in widely different individual and societal organizations, including insects and primates. In several species, the costs of altruistic behavior are measured in the loss of reproductive fitness, such as in insects. This behavior required a correction of Darwin's general theory of survival by replacing individual with colony survival.

In insects, the question is raised whether altruism is promoted by high relatedness as proposed by the inclusive fitness theory or when workers' behavior and status follow coercive interactions, as summarized in Ratnieks and Wenseleers (2007).

However, controversies arise when comparing altruistic or prosocial animal behavior with humans, as Silk and House (2016) and Gargagliano (2019) analyzed. The former authors place the core difference in terms of:

> Cooperation among unrelated individuals, who do not share direct genetic interests in offspring, is uncommon in nature but ubiquitous in human societies.

Analysis of altruistic or cooperative behaviors in nonhuman primates has been primarily performed in laboratory settings (i.e., artificial) (Vitale and Alleva 2007), which does not allow for a definite evaluation. One exception would be the observations performed by Seyfarth and Cheney (1984) in vervet monkeys (Old World monkeys of the family Cercopithecidae):

> Although the potential for reciprocal altruism exists in many animal societies, most interactions occur between closely related individuals, and examples of reciprocity among non-kin are rare[2,3]. The field experiments on vervet monkeys which we present here demonstrate that grooming between unrelated individuals increases the probability that they will subsequently attend to each other's solicitations for aid.

In a later article, Warneken et al. (2007) presented experimental evidence obtained in semi-free-ranging conditions:

> … chimpanzees act altruistically toward genetically unrelated conspecifics. In addition, in two comparative experiments, we found that both chimpanzees and human infants helped altruistically, regardless of any expectation of reward…

Authors conclude that the evolutionary roots of altruism among primates may thus reach as far back as the last common ancestor (*Pan*) of humans and chimpanzees.

In this regard, Schmelz et al. (2017), based on their observations on chimpanzees' behavior in experimental conditions, reported that:

> …findings demonstrate experimentally that chimpanzees are willing to incur a material cost to deliver rewards to a conspecific, but only if that conspecific previously assisted them, and particularly when this assistance was risky. Some key motivations involved in human cooperation thus may have deeper phylogenetic roots than previously suspected.

[5] Altruistic acts are those that intentionally benefit another organism, incur no direct personal benefit, and sometimes bear a personal cost. De Quervain et al. 2004. The *Stanford Encyclopedia of Philosophy* («Biological Altruism», 2003) characterizes biological altruism in terms of the behavior that benefits other organisms, placing at risk its own condition. According to the Dictionary of the Spanish Royal Academy, it represents "the drive to seek third party wellbeing at risk of its own."

Besides the fact that field observations may provide different data on nonhuman primate behavior, differences with the human species are based on the comparative cultural complexities. This assigns values and altruistic behaviors to a broader set of circumstances (mediated by symbolic and phonetic language) and levels of relatedness.

Regarding possible neurobiological substrates associated with differences in social behaviors, it is interesting to mention comparative studies performed in bonobos and chimpanzees (Issa et al. 2019). Species-specific social behavior was described in these two subspecies of the *Pan* order. While bonobos exhibit a greater social tolerance and higher sociosexual interactions, chimpanzees are more aggressive, territorial, and more significant risk-takers. According to the reported study, their brains differ in the microanatomical organization of circuits involved in socio-emotional systems.

In this context and regarding human (*conditional*) ultra-sociality, it is interesting to quote Jensen et al. (2014):

> Outsiders in a society of chimpanzees, for instance, would not expect to receive offers of food or solicitude; rather, they would be fiercely attacked. Even when well-integrated within a group, simple acts such as food sharing come only with begging and harassment (Stevens, 2004; Gilby, 2006). That is not to say that chimpanzees and other species do not engage in mutualistic, and sometimes coordinated, actions with one another. Social life is, for the most part, peaceful. But the fact that humans can interact in a peaceful, coordinated way, with a clear division of labor with unrelated individuals has earned our species the label (granted, a self-made label) of being ultrasocial (Szathmáry and Maynard Smith, 1995; Richerson and Boyd, 1998; Hill et al. 2009).

According to Choi (2011):

> 'The idea that chimps are indifferent to the welfare of others can now hopefully be put to rest,' de Waal stated.

It is a significant observation that chimpanzees (*Pan troglodytes*) return favors to conspecifics who have previously assisted them in acquiring food even at a material cost to themselves. According to the authors, chimpanzees can engage in materially costly reciprocal interactions (Fig. 4.1), a prosocial behavior commonly considered unique to humans. As Silk (2009) states:

> Had Darwin known that other primates distinguish kin from non-kin, form enduring relationships with their offspring, selectively groom, support and reconcile conflicts with their relatives and are aware of the kinship relationships between other group members, he would have been even more certain of the deep evolutionary links between humans and other animals.

Characteristics of prosocial behavior in humans have been singled out by several authors as having a species-specific profile and complexity (e.g., reviewed in Jensen et al. 2014), as synthetically stated:

> Different underlying causes can lead to similar outcomes.

In this regard, it seems that the outcome provides a common evolutionary context to a given species behavioral profile. Hence, available information across the animal

Fig. 4.1 Evolution of "power politics." (Shutterstock.com)

kingdom suggests that similar overt social behaviors reflect an evolutionary link, provided that ecological domains and species-specific adaptations are as diverse as animal orders and species exist. We, humans, attach additional cultural complexities, based on virtual and instrumental (technological) dimensions.

As predicted by cultural evolution and interdependence models, prosocial behavior in mammalian species is highly influenced by social norms and "political strategies" and alliances to cope with power distribution. Perhaps the observations made in honeybees provide what would seem to represent a universal "social unrest" reformulating power distribution, whether provided by internal or external means. Moritz et al. (2011) provided an example of "social power unrest" in insects:

> In honeybees, Apis mellifera, a suite of mechanisms, including physical dominance and pheromonal signals as well as policing behavior of workers regulate the reproductive hierarchies suppressing major intracolonial conflict among the workers. Whenever queen control is relaxed, workers can activate their ovaries and produce own offspring.

According to Jensen et al. (2014), in humans, motivational, emotional, and normative substrates of prosociality developed in childhood and emerged in our evolutionary history. As it will be discussed in the following chapters, deviant cultural components that excel in modern human societies—profit, greediness, and social prejudice—have often twisted and negated the mechanisms of prosocial behavior, as described by Jensen et al. (2014). These components of human society relativize the universal condition of social behaviors based on cooperation and solidarity.

Physical and emotional dominance/prevalence and punishment, or its menace, contribute to cooperative and submissive behaviors. In humans, although analysis of altruistic behavior would require a more complex analysis as proposed below, according to Ratnieks and Wenseleers (2007):

> Overall, it is clear, however, that in humans, punishment forms a powerful mechanism for promoting cooperation. In an experimental setting, the presence of punishment increased levels of cooperation [67]. In a comparative study of 15 small-scale societies, levels of

> cooperation were correlated with the degree to which uncooperative individuals were punished [68]. This finding parallel results from insect societies showing that more stringent policing promotes greater worker altruism.

As discussed in Colombo (2013), other values besides kin proximity trigger altruistic behavior at the human level. It tends to reduce or eliminate potential or actual conflict between the motivation to act according to given social, moral, or religious values adopted as own and the generated fear of not behaving accordingly. Altruistic behavior tends to relax tension in this intimate's conflict. Its resolution brings the feeling of "pleasure" of acting according to "superior rules." Hence, does this imply a hedonic behavior? If so, moral structure enforced by belonging and abiding by group, religious, political, or traditional constructs becomes an instrument for one fundamental biological component, such as satisfying hedonic drives. In this case, what gives an altruistic character to such behavior depends on the third party's valuation, internalized in the acting individual. Hence, is the pleasure of abiding by shared—culturally imposed—moral values or plain, autonomous, solidarity, and social responsibility decided by social influences or genetics?

On this line of thought, it is interesting to note that neural circuits activated by "altruistic" behavior overlap circuits involved in the mental processing of situations associated with pleasure that follows rewards. These circuits involve the insula, dorsal, and ventral striatum (de Quervain et al. 2004; Harbaugh et al. 2007; Carlson et al. 2011). These observations place altruistic behavior among the spectrum of hedonic drives, the search for reducing tension between actual and expected behavior. Performing dangerous acts in favor of third parties or kin relatives implies that the expected conscience of moral reward overpowers personal fate. Hence, the concept of reward appears to be tightly related to altruistic behavior and suggests that the reward is at the base of most behaviors. This issue projects to the following social domain, i.e., on which values do we place reward feelings? As mentioned above, the neuroscientific substrate involves neural mesencephalic-cortical-limbic circuits, which could be culturally programmed or biased, aimed toward different sets of social values and individual behavioral priorities.

This feeling of reward may be hierarchically imposed—avoiding punishment in hierarchical structures such as in religious domains and autocratic organizations—or built through experience domains. Physical or virtual (imagined) punishment has been implied in the process of performing altruistic behaviors, as mentioned by Buckholtz and Marois (2012):

> ... large-scale cooperation among unrelated individuals—a defining signature of H. sapiens culture—is predicated on our unique ability to establish norms, to transmit these norms from generation to generation and to enforce these norms through punishment.

Besides the above arguments, a game based on neuroeconomics allows to explore neural brain correlates during the act of accepting or rejecting economic proposals under conditions of definite game rules. Fehr and Gachter (2002) analyzed the underline role of punishment in altruistic decisions. To do so, they applied the ultimatum game[6] as originally proposed by Guth et al. (1982) and concluded that the introduction or elimination of the punishment opportunity led to an immediate raise, or fall, in the investing game. These observations are consistent with the concept that emotions are an essential factor behind altruistic punishment. Sanfey et al. (2003) observed while monitoring functional magnetic resonance on the players that unfair offers elicited neural activity in brain areas related to emotion and cognition (anterior insula and dorsolateral prefrontal cortex, respectively).

Knoch et al. (2006), also utilizing the ultimatum game and applying low-frequency transcranial magnetic stimulation—which leads to a suppression of activity in the stimulated brain region—explored how the human brain limits selfish motives and decides appropriate behavior. It was assumed that the dorsolateral prefrontal cortex (DLPFC) would be part of a network that modulates this behavior. This procedure revealed that disruption of the right DLPFC reduces the willingness to reject unfair offers, making them less able to resist the economic temptation to accept these offers. These data underline the relationship between the organization of specific brain circuits and human behavior components.

Threat Assessment Neural Systems and Inclusive Fitness

An interesting proposition mentioned by Flannery et al. (2007) links awareness of potential harm or one's sense of vulnerability with evolutionary threat assessment neural systems, involving the prefrontal cortex, basal ganglia, and circuits within the limbic system. The mentioned emotional dimensions support the development of behavioral constructs and mechanisms (rituals, feelings, religiosity) aimed at buffering them. In 1964, Hamilton developed a genetic mathematical model which allows for interactions between relatives on one another's fitness. According to the author, this model elaborates on the concept of *inclusive fitness*, which states that:

> Species following the model should tend to evolve behaviour such that each organism appears to be attempting to maximize its inclusive fitness. This implies a limited restraint on selfish competitive behaviour and possibility of limited self-sacrifices.

According to this theory, an organism's genetic success would be derived from cooperation and altruistic behavior. In other words, altruism among organisms which share a given percentage of genes enables these to be passed on to subsequent generations. Consequently, an altruistic act that supports the survival of other individuals theoretically enhances the genetic fitness of both the recipient and the altruistic organism. Or, as expressed by Bourke (2014):

[6] The *ultimatum game* consists essentially of a game between two operators, one that proposes and one that responds to the offer of the first. There is a fixed sum of money at stake, which the former offers—or not—to share with the second. If the latter rejected the offer, they'd both stop making money.

> … inclusive fitness theory finds that selection on a gene for social behaviour is determined by the gene's effects not only on the direct fitness of the bearer but also on the direct fitness of other individuals bearing the same gene (co-bearers, usually relatives) affected by the behavior.

Nowak et al. (2010) challenge this theory based on the alleged number of stringent assumptions; additionally, it would not provide further biological insight, as claimed by Abbot et al. (2011).

Silk and House (2016) pose the emergence of the capacity for large-scale cooperation and altruistic social preferences among humans within three possible domains, which belong to an ongoing debate. These are standard evolutionary foundations (as cooperation expressed in other animal societies) with shared fundamental elements of social preferences, selective evolutive pressures favoring cooperative breeding, or derived capacities for collaboration and group-level cooperation. An additional perspective belongs to the neuroscientific field regarding neural circuits involved, as briefly described.

Population Crowding and Altruistic Behavior

An additional factor in the emergence or practice of altruistic behavior consists of the consequences of population growth.

Altruistic behavior is dependent on external variables such as overcrowding and food accessibility, as demonstrated in rats by Calhoun (1962a, b, 1973), as is further described below. The mentioned variables may promote pathologic or aberrant behaviors that override altruism.

In addition to species-specific sets of behavioral profiles and social structure and interactions, population growth—besides other implications of the demographic analysis—represents a source for dynamic developments in social structure and fate. Schematically, depending on the considered species, a few alternatives to decrease social entropy ensue with increasing population densities. Besides egalitarian communities, one is the emergence of hierarchical structures; two, the process of fission generating migratory groups; and three, a potentially degrading social process, as reflected in the experimental setups by Calhoun (1962a, b, 1973). The consequences of overpopulation in these experiments (under spatially limited conditions, but with free-feeding access) are summarized in the following quotations (see also Chap. 8):

> When a population of laboratory rats is allowed to increase in a confined space, the rats develop acutely abnormal patterns of behavior that can even lead to the extinction of the population. (Calhoun 1962a)

> Once the number of rats in a room increased above a certain level, this frequency of contact while eating increased sufficiently that the rats developed a new definition of the feeding situation to include the presence of another rat. Gradually eating in the other three pens declined until 60-80 per cent of all food consumption was in these one of the four pens. The

> development of this atypical aggregation under the influence of the several conditions and processes involved forms what I have termed a *behavioral sink*.
>
> On the behavioral side, males developed a pansexuality in which they would mount other rats regardless of their age, sex, or receptivity.
>
> ... the development of a behavioral sink leads to a state of sustained inordinate aggregation which may be called 'pathological togetherness.' (Calhoun 1962b)

> If opportunities for role fulfilment fall far short of the demand by those capable of filling roles, and having expectancies to do so, only violence and disruption of social organization can follow. Individuals born under these circumstances will be so out of touch with reality as to be incapable even of alienation. Their most complex behaviours will become fragmented. Acquisition, creation and utilization of ideas appropriate for life in a post-industrial cultural-conceptual technological society will have been blocked. (Calhoun 1973)

Various authors have discussed the incidence and psychological effects of population density and crowding. In this context, Proshansky's views (1984) may be summarized regarding individual psychological processes and functions (stressful conditions) and physical settings, emphasizing personal space, human privacy, and human territoriality. The latter is an ancient, basic component in the evolution of human behavior, as mentioned earlier.

As Choldin (1978) would clearly state it and further enriched by Calhoun's animal research:

> In the mid-1970s there was considerable interest in sociology and psychology in urban population density and crowding. The interest centered about the assertion that high density generated social pathologies of several varieties. I refer to this assertion as the density-pathology hypothesis. (Choldin 1978)

Finally, to close this chapter, it seems opportune to quote Harbaugh et al. (2007) on the hedonic component that may transpire when altruism is applied to its instrumental use in political decisions that attempts to gather supporters either in populist strategies or when pursuing common good:

> Both pure *altruism* (satisfied by increases in the public good no matter the source or intent) and *warm-glow* (only fulfilled by an individual's own voluntary donations) motives appear to determine the hedonic consequences of financial transfers to the public good.

This poses an additional insight into social measures confronting deviant human behavior if we accept that cooperation and the punishment of defectors activate reward-related neural circuits. According to Fehr and Rockenbach (2004), this would suggest that evolution has endowed humans with proximate mechanisms that render altruistic behavior psychologically rewarding. It remains unsettled whether this represents a genuinely evolutionary acquisition or a spinoff subjected to cultural dynamics—thus, a conditioned behavior.

References

Abbot P, Abe J, Alcock J, Alizon S, et al. Inclusive fitness theory and eusociality. Nature. 2011;471:E1–4. https://doi.org/10.1038/nature09831.

Altman J, Das GD. Autoradiographic examination of the effects of enriched environment on the rate of glial multiplication in the adult rat brain. Nature. 1964;204:1161–3.

Alward BA, Laud VA, Skalnik CJ, York RA, Juntti SA, Fernald RD. Modular genetic control of social status in a cichlid fish. PNAS. 2020;117:28167–74. https://doi.org/10.1073/pnas.2008925117.

Amrein I, Becker AS, Engler S, Huang S-h, Müller J, Slomianka L, Oosthuizen MK. Adult neurogenesis and its anatomical context in the hippocampus of three mole-rat's species. Front Neuroanat. 2014;8:1–11. https://doi.org/10.3389/fnana.2014.00039.

Atzil S, Gao W, Fradkin I, Barrett LF. Growing a social brain. Nat Hum Behav. 2018;2:624–36.

Barchuk AR, Cristino AS, Kucharski R, Costa LF, Simões ZLP, Maleszka R. Molecular determinants of caste differentiation in the highly eusocial honeybee *Apis mellifera*. BMC Dev Biol. 2007;7:70. https://doi.org/10.1186/1471-213X-7-70.

Beery AK, Kaufer D. Stress, social behavior, and resilience: insights from rodents. Neurobiol Stress. 2015;1:116–27.

Beery AK, Holmes MM, Lee W, Curley JP. Stress in groups: lessons from non-traditional rodent species and housing models. Neurosci Biobehav Rev. 2020;113:354–72. https://doi.org/10.1016/j.neubiorev.2020.03.033.

Bourke AFG. Hamilton's rule and the causes of social evolution. Philos Trans R Soc Lond Ser B Biol Sci. 2014;369(1642):20130362. https://doi.org/10.1098/rstb.2013.0362.

Buckholtz JW, Marois R. The roots of modern justice: cognitive and neural foundations of social norms and their enforcement. Nat Neurosci. 2012;15:655–61.

Calhoun JB. Population density and social pathology. Scientific American, Inc. 1962a;140.142.214.159.

Calhoun JB. A behavioral sink, Chapter 22. In: Bliss EL, editor. From roots of behavior. 1962b.

Calhoun JB. Death squared. The explosive growth and demise of a mouse population. Proc R Soc Med. 1973;66(1 Pt 2):80–8.

Carlson JM, Foti D, Mujica-Parodi LR, Harmon-Jones E, Hajcak G. Ventral striatal and medial prefrontal BOLD activation is correlated with reward-related electrocortical activity: a combined ERP and fMRI study. NeuroImage. 2011;57:1608–16.

Choi CQ. Like humans, chimps show selfless behaviors. Live Sci. 2011.

Choldin HM. Urban density and social pathology. Annu Rev Sociol. 1978;4:91–113.

Colombo JA. Bajo libertad condicionada. (Under conditioned freedom). Ed. Imago Mundi; 2013.

Colombo JA. Our Animal condition and social construction. 2019. Nova Sci. Publ., USA. ISBN: 978-1-53615-357-6. eBook ISBN: 978-1-53615-3583.

De Quervain D, Fischbacher U, Treyer, Schellhammer M, Schnyder U, Buck A, Fehr E. The neural basis of altruistic punishment. Science. 2004;305:1254–8.

Eriksson PS, Perfilieva E, Björk-Eriksson T, Alborn AM, Nordborg C, Peterson DA, Gage FH. Neurogenesis in the adult human hippocampus. Nat Med. 1998;4:1313–7.

Eshel I, Feldman MW. Individual selection and altruistic relationships: the legacy of W. D. Hamilton. Theor Popul Biol. 2001; https://doi.org/10.1006/tpbi.2000.1507.

Fang W-Q, Yuste R. Overproduction of neurons is correlated with enhanced cortical ensembles and increased perceptual discrimination. Cell Rep. 2017;21:381–92. https://doi.org/10.1016/j.celrep.2017.09.040.

Fehr E, Gachter S. Altruistic punishment in humans. Nature. 2002;415:137–40. https://doi.org/10.1016/j.conb.2004.10.007.

Flannery DJ, Singer MI, van Dulmen M, Kretschmar JM, Belliston LM. Exposure to violence, mental health, and violent behavior. In: Flannery DJ, Vazsonyi AT, Waldman ID, editors. The Cambridge handbook of violent behavior and aggression. New York: Cambridge University Press; 2007. p. 306–21. https://doi.org/10.1017/CBO9780511816840.015.

Gargagliano A. Altruistic behavior in primates. Aisthesis. 2019;10:47–56.

Gheusi G, Ortega-Perez I, Murray K, Lledo P-M. A niche for adult neurogenesis in social behavior. Behav Brain Res. 2009;200:315–22.

Gould E, McEwen BS, Tanapat P, Galea LAM, Fuchs E. Neurogenesis in the dentate gyrus of the adult tree shrew is regulated by psychosocial stress and NMDA receptor activation. J Neurosci. 1997;17:2492–8.

Gould E, Tanapat P, McEwen BS, Flugge G, Fuchs E. Proliferation of granule cell precursors in the dentate gyrus of adult monkeys is diminished by stress. Proc Natl Acad Sci U S A. 1998;95:3168–71.

Güth W, Schmittberger R, Schwarze B. An experimental analysis of ultimatum bargaining. J Econ Behav Organ. 1982;3:367–88.

Hackman DA, Farah MJ. Socioeconomic status and the developing brain. Trends Cogn Sci. 2009;13:65–73.

Hackman DA, Farah MJ, Meaney MJ. Socioeconomic status and the brain: mechanistic insights from human and animal research. Nat Rev Neurosci. 2010;11:651–9. https://doi.org/10.1038/nrn2897.

Hamilton WD. The genetical evolution of social behavior. I J Theor Biol. 1964;7:1–16.

Harbaugh WT, Mayr U, Burghart DR. Neural responses to taxation and voluntary giving reveal motives for charitable donations. Science. 2007;316:1622–5.

Harlow HF. Love in infant monkeys. In: Psychobiology, vol. 200. San Francisco: Scientific American, W.H. Freeman, and Co; 1959. p. 100–6.

Henrich J, Boyd R. Division of labor, economic specialization, and the evolution of social stratification. Curr Anthropol. 2008;49:71724.

Holmes MM, Rosen GJ, Jordan CL, de Vries GJ, Goldman BD, Forger NG. Social control of brain morphology in a eusocial mammal. PNAS. 2007;104:10548–52.https://doi.org/10.1073/pnas.0610344104.

Issa FA, Drummond J, Cattaert D, Edwards DH. Neural circuit reconfiguration by social status. J Neurosci. 2012;32:5638–45.

Issa HA, Staes N, Diggs-Galligan S, Stimpson CD, Gendron-Fitzpatrick A, Taglialatela JP, Hof PR, Hopkins WD, Sherwood CC. Comparison of bonobo and chimpanzee brain microstructure reveals differences in socio-emotional circuits. Brain Struct Funct. 2019;224:239–51. https://doi.org/10.1007/s00429-018-1751-9.

Jensen K, Vaish A, Schmidt MFH. The emergence of human prosociality: aligning with others through feelings, concerns, and norms. Front Psychol. 2014;5:1–16.

Jones BM, Rao VD, Gernat T, Jagla T, Cash-Ahmed AC, Rubin BER, Comi TJ, Bhogale S, Husain SS, et al. Individual differences in honeybee behavior enabled by plasticity in brain gene regulatory networks. eLife Ecol Genet Genom. 2020; https://doi.org/10.7554/eLife.62850.

Knoch D, Pascual-Leone A, Meyer K, Treyer V, Fehr E. Diminishing reciprocal fairness by disrupting the right prefrontal cortex. Science. 2006;314:829–32.

Kozorovitskiy Y, Gould E. Dominance hierarchy influences adult neurogenesis in the dentate gyrus. J Neurosci. 2004;24:6755–9.

Krech D, Rosenzweig MR, Bennett EL. J Comp Physiol Psychol. 1960;53:509–14.

Lieberwirth C, Wang Z. The social environment and neurogenesis in the adult mammalian brain. Front Hum Neurosci. 2012;6:1–19. https://doi.org/10.3389/fnhum.2012.00118.

Maruska KP, Zhang A, Neboori A, Fernald RD. Social opportunity causes rapid transcriptional changes in the social behavior network of the brain in an African cichlid fish. J Neuroendocrinol. 2012; https://doi.org/10.1111/j.1365-2826.2012.02382.x.

Michener CD, Brothers DJ. Were Workers of Eusocial Hymenoptera Initially Altriuistic or Oppressed? Proc. Nat. Acad. Sci. 1974:71:671–674. USA

Miller TH, Clements K, Ahn S, Park C, Ji EH, Issa FA. Social status -dependent shift in neural circuit activation affects decision making. J Neurosci. 2017;37:2137–48.

Moda LM, Vieira J, Guimarães Freire AC, Bonatti V, Bomtorin AV, Barchuk AG, Simões ZLP. Nutritionally driven differential gene expression leads to heterochronic brain development in honeybee castes. PLoS One. 2013;8(5):e64815. https://doi.org/10.1371/journal.pone.0064815.

Moritz RFA, Lattorff HMG, Crous KL, Hepburn RH. Social parasitism of queens and workers in the Cape honeybee (*Apis mellifera capensis*). Behav Ecol Sociobiol. 2011;65:735–40. https://doi.org/10.1007/s00265-010-1077-y.

Nilsson M, Perlieva E, Johansson U, Orwar O, Eriksson PS. Enriched environment increases neurogenesis in the adult rat dentate gyrus and improves spatial memory. JNeurobiol. 1999;39:569–78.

Nowak MA, Tarnita CE, Wilson EO. The evolution of eusociality. Nature. 2010;466:1057–62. https://doi.org/10.1038/nature09205.

O'Riain MJ, Jarvis JUM, Alexander R, Buffenstein R, Peeters C. Morphological castes in a vertebrate. PNAS. 2000;97:13194–7. https://doi.org/10.1073/pnas.97.24.13194.

Pahlke S, Jaumann S, Seid MA, Smith AR. Brain differences between social castes precede group formation in a primitively eusocial bee. Sci Nat. 2019;106:49. https://doi.org/10.1007/s00114-019-1644-7.

Proshansky H. Population change and human conflict. Chapter 3. In: Choucri N, editor. Multidisciplinary perspective on population and conflict. New York: Syracuse University Press; 1984. p. 59–90.

Ratnieks FLW, Wenseleers T. Altruism in insect societies and beyond: voluntary or enforced? Trends Ecol Evol. 2007;23:45–52. https://doi.org/10.1016/j.tree.2007.09.013.

Sanfey AG, Rilling JK, Aronson JA, Nystrom LE, Cohen JD. Decision-making in the ultimatum game. Science. 2003;300:1755–8. https://doi.org/10.1126/science.1082976.

Scantlebury M, Speakman JR, Oosthuizen MK, Roper TJ, Bennett NC. Energetics reveals physiologically distinct castes in a eusocial mammal. Nature. 2006; 440:795–797. http:10.1038/nature04578.

Schmelz M, Grueneisen S, Kabalak A, Jost J, Tomasello M. Chimpanzees return favors at a personal cost. PNAS. 2017;114:74627467.

Seyfarth RM, Cheney DL. Grooming, alliances, and reciprocal altruism in vervet monkey. Nature. 1984;308:541–3.

Sherman PW, Lacey EA, Reeve HK, Keller L. The eusociality continuum. Behav Ecol. 1995;6:102–8.

Shultz S, Dunbar R. Encephalization is not a universal macro evolutionary phenomenon in mammals but is associated with sociality. PNAS. 2010;107:21582–6.

Silk JB. Nepotistic cooperation in nonhuman primate groups. Philos Trans R Soc B. 2009;364:3243–54. https://doi.org/10.1098/rstb.2009.0118.

Silk JB, House BR. The evolution of altruistic social preferences in human groups. Philos Trans R Soc B. 2016;371:20150097. https://doi.org/10.1098/rstb.2015.0097.

Turner AM, Greenough WT. Differential rearing effects on rat visual cortex synapses. Synaptic and neuronal density and synapses per neuron. Brain Res. 1985;329:195–203.

Uylings HBM, Kuypers K, Veltman WAM. Environmental influences on the neocortex in later life. Prog Brain Res. 1978;48:261–74.

Vitale A, Alleva E. Altruism in human and non-human animals. J Anthropol Sci. 2007;85:237–9.

Volkmar FR, Greenough WT. Rearing complexity affects branching of dendrites in the visual cortex of the rat. Science. 1972;176:1445–7.

Warneken F, Hare B, Melis AP, Hanus D, Tomasello M. Spontaneous altruism by chimpanzees and children. PLoS Biol. 2007.

Weiner SA, Toth AL. Epigenetics in social insects: a new direction for understanding the evolution of castes. Genet Res Int. 2012;Article ID 609810. https://doi.org/10.1155/2012/609810.

Whitehead H, Laland KL, Rendell L, Thorogood R, Whiten A. The reach of gene-culture coevolution in animals. Nat Commun. 2019;10:2405. https://doi.org/10.1038/s41467-019-10293-y.

Yamamuro K, Bicks LK, Leventhal MB, Kato D, Im S, Flanigan ME, Garkun Y, Norman KJ, Caro K, Sadahiro M, Kullander K, Akbarian S, Russo SJ, Morishita H. A prefrontal -paraventricular thalamus circuit requires juvenile social experience to regulate adult sociability in mice. Nat Neurosci. 2020;23:1240–52.

Yamazakia Y, Shiraia K, Paul RK, Fujiyuki T, Wakamoto A, Takeuchi H, Kubo T. Differential expression of HR38 in the mushroom bodies of the honeybee brain depends on the caste and division of labor. FEBS Lett. 2006;580:2667–70.

Yap JJ, Takase LF, Kochman LJ, Fornal CA, Miczek KA, Jacobs BL. Repeated brief social defeat episodes in mice: effects on cell proliferation in the dentate gyrus. Behav Brain Res. 2006;172:344350.

Zayed A, Robinson GE. Understanding the relationship between brain gene expression and social behavior: lessons from the honeybee. Annu Rev Genet. 2012;46:591–615. https://doi.org/10.1146/annurev-genet-110711-15551.

Chapter 5
Social Inequality

> ...definitional problems (*of eusociality*) could be reduced by expanding the eusociality concept to include all vertebrate and invertebrate societies with helpers. We favor the latter approach. It seems more productive to recognize that similar social systems occur in birds, mammals, and insects than to debate whether particular insects are eusocial (e.g., Furey, 1992 versus Tsuji, 1992). (Sherman et al. 1995) (text in italics, added by JAC)

> In eusocial organisms, some individuals specialize in reproduction and others in altruistic helping. The evolution of eusociality is, therefore, also the evolution of remarkable inequality. (Ratnieks and Helantera 2009)

> Human society operates within this didactic tension between dominance and equality, between hierarchical and egalitarian, between modes of behavior that feature or privilege the group to those that accent individuals. (Price and Feinman 2010)

Inequality in Social Construction

It would seem a necessary introductory requisite on the issue of cooperative breeding to quote C. Darwin (1874) on this issue:

> As the social instincts both of man and the lower animals have no doubt been developed by nearly the same steps, it would be advisable, if found practicable, to use the same definition in both cases, and to take the standard of morality, the general good or welfare of the community, rather than the general happiness...

Also, as Silk (2009) comments:

> Darwin was struck by the many similarities between humans and other primates and believed that these similarities were the product of common ancestry. He would be even more impressed by the similarities if he had known what we have learned about primates over the last 50 years. Genetic kinship has emerged as the primary organizing force in the evolution of primate social organization and the patterning of social I in non-human primate groups.

J. A. Colombo, *Dominance Behavior*, https://doi.org/10.1007/978-3-030-97401-5_5

According to Hayden and Villeneuve (2010):

> There is a major rift among archaeological theorists as to whether political complexity and socioeconomic inequality emerged to serve the common good especially in enhancing defense, production, and distribution (Diehl 2000, Johnson 1982, Pebbles and Kus 1977, Saitta and Keene 1990:213–214, Saitta 1999, Suttles 1968), whether they emerged as a means of promoting elite self-interests or even exploitation by leaders (Diehl 2000, Earle 1977, 1978, 1997, Gilman 1981, Roscoe 2000), or whether they emerged from religious beliefs or other cultural values.

Concepts involved in the following statement by Hardisty and Cassill (2010) merge with the inclusiveness expressed by Nowak (2006) and Nowak et al. (2010)—as mentioned earlier—involving human social structures:

> Eusociality is a term that describes social invertebrates such as ants, bees, wasps, termites, some aphids, some thrips (Batra 1966; Wilson 1971; Crespi 1992; Stern and Foster 1996) and social shrimps (Duffy et al. 2000; Duffy 2003). Three criteria describe eusociality: overlapping generations in which parents, offspring and siblings all reside in one domicile or territory; cooperative rearing of young by individuals other than a parent; and a reproductive division of labor with fertile breeders and sterile helpers (Wilson 1971; Foster and Ratnieks 2005).

As described previously, cooperative breeding (conspecifics helping parents to raise their young) is a characteristic among most social insects, birds, and mammals. Whether it applies to human beings is under discussion, as Burkart et al. (2017) state that our cooperative breeding would have arisen from a different ancestral state than in other species. Besides its hypothetical evolutionary origin and operational profiles in social groups, cooperative breeding represents an evolutionary widespread social behavior, as it also does social inequality. Henrich and Boyd (2008) developed a mathematical model showing how economic inequality between social groups can arise with the critical assumption that human populations structure into groups and that cultural learning is more likely to occur within than between groups. Stratification would eventually take place if there were potential gains from specialization and exchange. According to these authors:

> This (*mathematical*) model predicts that stratification is favored, *ceteris paribus* (*meaning "all other things being equal"*), by (1) greater surplus production, (2) more equitable divisions of the surplus among specialists, (3) greater cultural isolation among subpopulations within a society, and (4) more weight given to economic success by cultural learners. (Words within parenthesis, in italics, added by JAC)

On the evolutionary domain, Richerson and Boyd (2005) argued further that human behavior would express:

> … two sets of innate social predilections: an older set that is part of our primate heritage and a newer set they call the "tribal instincts" that evolved as part of biological-cultural convolution. Among the ancient instincts is the quest for status; among the tribal instincts are cooperation in large groups, and reciprocity.

This comment brings up *egalitarianism* among humans, the origin and evolution of which represents a theme of discussion among anthropologists. Ames (2010) proposes a general view in the following paragraph:

> Perhaps over much of the last 100,000 years, the majority of human societies were small, with prestige competition and fluid rank orders, with a minority of formally egalitarian

> societies and another minority with stable, sometimes materially visible rank orders or perhaps even formal ranking. Over the relatively short spans of centuries, societies shifted back and forth across these social forms; over millennia, they went extinct, giving the archaeological record of the last 100,000 years or so its flickering character.

This theme considers several central aspects of human behavior, such as moral issues, hierarchies and means in social organization, and censorship. According to Boehm (1993, 2012a, b), egalitarianism does not exist among hunter-gatherers. Instead, he proposes a "reverse dominance hierarchy" in which coalitions of subdominant individuals usually prevent alpha individuals from exercising dominance. In terms of interactions, modern human foragers have the potential for full-scale war. Although hunter-gatherers can resolve conflicts preemptively, they also use mechanisms such as truces and peace pacts to mitigate conflict when the costs become too high. Boehm (2012b) places the issue on a comparative evolutive domain when referring to the ancestral *Pan*, a predecessor of *Homo*, bonobos, and chimpanzees living in social dominance hierarchies. This would have created conflict through individual and coalitional competition. Extant hunter-gatherers would have suppressed dominance and competition (except on reproductive issues).

Besides its significance in scientific analysis, these considerations also draw attention to our general theme of ancient drives. As mentioned earlier, the social drift from castes to classes to prejudices in modern societies is a component of cultural construction in its broader sense.

In terms of comparative evolution, these social components draw a horizon of unequal rights among social species. This widespread operative concept represents a primary significance to be considered in terms of social changes among humans. How can our species avoid evolutive inertias or pressures that drift toward forms of social inequality imposed by seemingly universal evolutive inertias on our animal construction?

Let us consider the basic statement that culture represents the product of social behavior transmissible through learning and potentially may affect or interact with the genetic expression of heritage. Should we aim to deconstruct the system based on inequalities in access to knowledge, creativity, and maximizing cognitive freedom, which would be the road to transit and the formal construction of such a society? Should it be based on a competitive basis or on cooperation?

The Concept of Castes Revisited

Although this theme has been previously discussed, as we deal with more complex (human) societies, it seems worth reinstating some basic definitions and drawing comparisons with concepts involving non-primate animals.

Applied to human social construction, the concept of caste has been approached by the Spanish Royal Academy (Real Academia de la Lengua).[1] It would imply the following:

[1] https://dle.rae.es/casta#7qBrUKC

> ...group that constitutes a special class that tends to remain separated due to race, religion or other social domain.

As defined by the *American Heritage Dictionary*:[2]

> ...an endogamous and hereditary social group limited to persons of the same rank, occupation, and economic position.

According to the New World Encyclopedia:[3]

> Anthropologists use the term more generally, to refer to a social group that is endogamous and occupationally specialized. Such groups are common in societies with a low degree of social mobility. In its broadest sense, examples of caste-based societies include colonial Latin America under Spanish and Portuguese rule, Japan, Korea, some parts of Africa, as well as across the Indian subcontinent.

Besides the historical representation of social stratification of prerevolutionary France, the illustration in Fig. 5.1 also depicts a metaphoric representation of an ongoing social stratification in most contemporary nations, despite variable improvement in social mobility. Class prejudices, hierarchical privileges, corporative financial empowerment, a gap in cognitive development, poverty, and social marginalization describe a modified picture of social stratification from the one depicted in Fig. 5.1, but not less questionable on equal rights and humanitarian terms.

Berreman (1967) defines a caste system when:

> ... a society is made up of birth-ascribed groups which are hierarchically ordered and culturally distinct. The hierarchy entails differential evaluation, rewards and association...

As it was commented in previous pages, the concept of caste applied to human societies is equivalent to the one described in other species. Its expression in different cultures draws close parallelisms and suggests an evolutionary continuity of its formal expression.

Martinez and Iyer (2008) place the issue of the racial caste system in Latin America in the following order:

> Latin American societies are structured in a fuzzy racial caste system. Whites are at the top. Next are individuals of mixed European and Indian or mixed European and African ancestry, the Mestizos and the Mulatos, respectively. At the bottom are the Indians and the blacks.

While this general statement would probably require further refinement, it seems unavoidable to draw the picture of social inequalities under an evolutive anthropological perspective—that is, the cultural expression of social inequalities under forms adapted to modern cultural development, such that caste, class, and prejudice represent progressively smoothed versions of inequality promoters.

It is interesting to include Mexico in the history of inequality, another historically significant Latin-American country in terms of native cultural development that underwent colonization by Spanish conquerors. There was close parallelism in social structure with other Spanish colonies on the American continent. They included the former Viceroyalty of the Río de la Plata or Virreinato del Rio de la

[2] https://www.ahdictionary.com/word/search.html?q=caste

[3] https://www.newworldencyclopedia.org/entry/caste_system

Fig. 5.1 In France before the French Revolution, society was divided into three estates: the clergy, nobility, and commoners. (Color modified, from https://socialsci.libretexts.org/Bookshelves/Sociology/Book%3A_Sociology_(Boundless)/08%3A_Global_Stratification_and_Inequality/8.01%3A_Systems_of_Stratification/8.1C%3A_Caste_Systems)

Plata, now roughly including the present-day territories of Argentina, Chile, Bolivia, Paraguay, and Uruguay. They were essentially composed of four levels as described by Martinez and de la Torre (2003). It consisted of the Spaniards at the top level, followed by Creoles (descendants of Spaniards but born in American continent colonies, with less presence in government affairs); both owned land and mines and occupied all dominant government positions. In third place came those who had Indian and Spanish ancestors, the mestizos, and at the bottom were the Indians, many of them slaves and poor. Manual labor was reserved for the Indians and mestizos. According to Martinez and de la Torre:[4]

> Yet nowadays the rich are almost exclusively white, they form what some call the Creole oligarchy. Mestizos are predominant in the small middle and working classes, while the poor are primarily Indians and Mestizos of marked Indian ancestry. That is, in 2002

[4] http://www.zonalatina.com/Zldata55.htm 2007

> Mexico's caste system, while not as rigid as in colonial times, it is still very much in operation. The existence of Mexico's resilient caste system is denied by most Mexicans and ignored in the literature about the cultural traditions that foreign firms face in this country.

Caste systems were present in several countries and persist nowadays under modified (feudal) social configuration that gave rise to outcastes or undercasts, as observed in India (whether Muslims or Christians), as well as in Bali, Pakistan, Nepal, and to some extent in Japan and several African countries. In ancient times, as in Egypt, this social stratification, with its cultural characteristics, was also present:

> The society of ancient Egypt was strictly divided into a hierarchy with the king at the top and then his vizier, the members of his court, priests and scribes, regional governors (eventually called 'nomarchs'), the generals of the military (after the period of the New Kingdom, c. 1570- c. 1069 BCE), artists and craftspeople, government overseers of worksites (supervisors), the peasant farmers, and slaves. Social mobility was not encouraged, nor was it observed for most of Egypt's history, as it was thought that the gods had decreed the most perfect social order which was in keeping with the central value of the culture…[5]

According to Berreman (1967):

> Manusmriti, widely regarded to be the most important and authoritative book on Hindu law and dating back to at least 1,000 years before Christ was born, 'acknowledges and justifies the caste system as the basis of order and regularity of society'.

Eusociality and Egalitarianism. Social Fitness and Dominance. From Castes to Classes, to Prejudice

In previous pages, the concept of eusociality as applied to insects and a short set of mammalian species was considered. Authors stress the impact of social hierarchies and early feeding on the development of social castes, indicating phenotypic plasticity as described by Weiner and Toth (2012):

> The extreme phenotypic plasticity of social insect castes has become even more compelling with the knowledge that, in most species, queen and worker caste differences are environmentally, not genetically, determined. With some notable exceptions, such as some genera of ants (e.g., [11–15]), in most social insects, there are no heritable genetic differences that dictate which individuals become queens and which become workers, nor among different morphological castes of workers [16].

However, these developmental conditioners are also found active in mammalian species besides the naked mole rat, though developing under different social and cultural settings. Perhaps the concepts of Lewis (1966) are pertinent to be mentioned in such context under the general title of *the culture of poverty*:

[5] [20]Ancient History Encyclopedia. Joshua J. Mark. https://www.ancient.eu/

> Does membership in a group that has been poor for generations constitute belonging to a separate culture? A study of Puerto Ricans in both Puerto Rico and New York indicates that it does.
>
> ...I have called the culture of poverty... In my writings it is the label for a specific conceptual model that describes in positive terms *a subculture of Western society with its own structure and rationale, a way of life handed on from generation to generation along family lines*. The culture of poverty is not just a matter of deprivation or disorganization, a term signifying the absence of something. It is a culture in the traditional anthropological sense in that it provides human beings with a design for living, with a ready-made set of solutions for human problems, and so serves a significant adaptive function. *This style of life transcends national boundaries and regional and rural-urban differences within nations. Wherever it occurs, its practitioners exhibit remarkable similarity in the structure of their families, in interpersonal relations, in spending habits, in their value systems and in their orientation in time*. (Italics by JAC)

The concepts of Lewis describing the culture of poverty closely resemble developmental conditions and generational persistence and hierarchical social conditioning previously described for eusocial species. Hence, the question should also be brought up regarding its impact on brain organization and cognitive and emotional development as previously discussed (Colombo 2007; Lipina and Colombo 2009; Colombo 2010). As mentioned by Weiner and Toth (2012), DNA methylation in mammalian species is critical in mediating the effects of early life nutrition and social circumstances on phenotype development, thus providing a molecular mechanism for the developmental impact of social inequalities.

Earlier anthropological studies by Davis et al. (1941) on the deep south of the United States underline the evolution of social structures in human civilization, traversing stages of caste organization, as described below:

> ...this study deals with the caste-class system in a typical urban-rural area in the deep South, which clearly divides the inhabitants into superordinate whites and subordinate Negroes. Within the castes are social classes. The class position of whites and Negroes within their respective color-castes establishes an appropriate norm of behavior which is accompanied by ideology and values different from those of persons in other positions. The class structure is mirrored in these ideologies just as the caste structure is expressed in caste dogmas. Both caste and class are hierarchies in which the lowest group symbolizes all the negative values of the society. They differ in that persons may sometimes change position in the class hierarchy; never in the caste hierarchy.

Both considerations made by Davis et al. (1941) and Lewis (1966) converge in a central point; the impact of sociocultural stratification on brain and mind development in modern societies. Although the dynamics of this sociobiological interaction in modern social economies are not identical to eusocial societies, the final impact on the brain and mind conserves worrying parallelisms. This is also described for current China social dynamics, as described by Dowe (2019):[6]

> ...the fruits of its success have yet to become available to everyone—and nowhere is this clearer than in the divide between urban and rural populations. China's system of hukou—home registration—has regimented the nation into two distinct and unequal castes. Today,

[6] China's Caste System Has Codified Economic Inequality. https://www.diplomaticourier.com/posts/chinas-caste-system-has-codified-economic-inequality. 2019

holders of rural hukou are ostracized and discriminated against by their fellow citizens, and have little choice but to take on the most dangerous, demanding, and low-paying jobs that those with urban hukou wouldn't dare involve themselves in.

This was achieved by classifying each Chinese citizen in a category —rural or urban— upon birth, and affording them certain privileges based on the distinction, with transfer between classes heavily limited. In practice, this has resulted in the exclusion of the rural class from social welfare programs, including subsidized housing, disability benefits, healthcare, and subsidized education.

These examples in dealing with human societies show that social stratification and its impact on the probabilities of individual development are not exclusive of the caste system reported in India. Regarding castes in India[7] and according to Archytas (2002),[8] the traditional caste system was like the ancient Egyptians, with the exception that in Egypt the pharaoh was at the head of the ruling and the religious class. In India, rulers were usually divided into a religious class (*brahman*) which had the monopole of the cultural knowledge and a dominant class (*kastriya*), with the monopole of the military and ruling powers.

Analysis of cultural development of beliefs and social institutions among the different human societies is well beyond the scope of this essay. The point to be made is that human societies have expressed basic social organizations reminiscent of several animal species mentioned previously through different epochs, including the present one. The chain of comparative evolution in this regard has a saltatory or discontinuous component in terms of complexity and conscient purpose. Yet, there is mounting evidence of universal components and drives in social construction among the animal kingdom, including humans. They include social stratification or inequalities in roles and benefits, punishment, power, and material and emotional dominance/prevalence (whether its origin recognizes religious or magical beliefs).

References

Ames KM. On the evolution of the human capacity for inequality and/or egalitarianism. Chapter 2. In: Price TD, Feinman GM, Springer C, editors. Pathways to power. New perspectives on the emergence of social inequality. New York: Science & Business Media LLC; 2010.

Berreman GD. Caste as social process. Southwest J Anthropol. 1967;23(4):351–70.

Boehm C. Egalitarian behavior and reverse dominance hierarchy. Curr Anthropol. 1993;34:227–40.

Boehm C. Costs and benefits in hunter-gatherer punishment. Behav Brain Sci. 2012a;35:19–20.

[7]What is India's caste system? https://www.bbc.com/news/world-asia-india-35650616. 2019. "Historians, though, say that until the 18th Century, the formal distinctions of caste were of limited importance to Indians, social identities were much more flexible, and people could move easily from one caste to another. New research shows that hard boundaries were set by British colonial rulers who made caste India's defining social feature when they used censuses to simplify the system, primarily to create a single society with a common law that could be easily governed."

[8]coexistencialism.tripod.com/caste-system/. Hector Archytas 19/11/2020. The difference between the ancient India caste system and medieval social classes

Boehm C. Ancestral hierarchy and conflict. Science. 2012b;336:844–7. https://doi.org/10.1126/science.1219961.

Burkart J, Schubiger M, Michèle N, van Schaik CP. The evolution of general intelligence. Behav Brain Sci. 2017;40:e195. https://doi.org/10.1017/S0140525X16000959.

Colombo JA. Pobreza y desarrollo infantil (Poverty and child development). Buenos Aires: Ed. Paidos; 2007.

Colombo JA. Somos la especie equivocada? (Are we the wrong species?). Buenos Aires: Ed. EUDEBA; 2010.

Darwin C. The descent of man and selection in relation to sex. Preface to the second edition. Chap. IV. 2nd ed. London; 1874. p. 121

Davis A, Gardner BB, Gardner MR. Deep South; a social anthropological study of caste and class. Chicago: University of Chicago Press; 1941. https://psycnet.apa.org/record/194200249-000

Hardisty BE, Cassill DL. Extending eusociality to include vertebrate family units. Biol Philos. 2010;25:437–40. https://doi.org/10.1007/s10539-009-9176-8.

Hayden B, Villeneuve S. Chapter 5. Who benefits from complexity? A view from futuna. In: Price TD, Feinman GM, editors. Pathways to power, fundamental issues in archaeology. Springer Science+Business Media, LLC; 2010. https://doi.org/10.1007/978-1-4419-63000,1,C. https://link.springer.com/bookseries/5972.

Henrich J, Boyd R. Division of labor, economic specialization, and the evolution of social stratification. Curr Anthropol. 2008;49:71724.

Lewis O. The culture of poverty. Sci Am. 1966;215:19–25. www.pnas.org/cgi/doi/10.1073/pnas.1620744114

Lipina SJ, Colombo JA. Poverty and brain development during childhood. Washington, DC: American Psychological Association; 2009.

Martinez R, de la Torre C. Mexicos resilient caste system: managerial implications for foreign firms. Int Bus Econ Res J (IBER). 2003;2(5) https://doi.org/10.19030/iber.v2i5.3802.

Martinez R, Iyer V. Latin America's racial caste system: salient marketing implications. Int Bus Econ Res J. 2008;7 https://doi.org/10.19030/iber.v7i11.3308.

Nowak MA. Five rules for the evolution of cooperation. Science. 2006;314:1560–3. https://doi.org/10.1126/science.1133755.

Nowak MA, Tarnita CE, Wilson EO. The evolution of eusociality. Nature. 2010;466:1057–62. https://doi.org/10.1038/nature09205.

Price TD, Feinman GM. Chapter 1. Social inequality and the evolution of human social organization. In: Price TD, Feinman GM, editors. Pathways to power, fundamental issues in archaeology. Springer Science+Business Media, LLC; 2010. https://doi.org/10.1007/978-14419-6300-0_1. https://link.springer.com/bookseries/5972.

Ratnieks FLW, Helantera H. The evolution of extreme altruism and inequality in insect societies. Philos Trans R Soc B. 2009;364:3169–79. https://doi.org/10.1098/rstb.2009.0129.

Richerson PJ, Boyd R. Not by genes alone. How culture transformed human evolution. Chicago/London: The University Chicago Press; 2005.

Sherman PW, Lacey EA, Reeve HK, Keller L. The eusociality continuum. Behav Ecol. 1995;6:102–8.

Silk JB. Nepotistic cooperation in nonhuman primate groups. Philos Trans R Soc B. 2009;364:3243–54. https://doi.org/10.1098/rstb.2009.0118.

Weiner SA, Toth AL. Epigenetics in social insects: a new direction for understanding the evolution of castes. Genet Res Int. 2012:Article ID 609810. Hindawi Publishing Corporation. https://doi.org/10.1155/2012/609810.

Chapter 6
Culture and Evolution

Darwin remarked both that 'there is no fundamental difference between man and the higher mammals in their mental faculties' (1871:35) and that 'the difference between the mind of the lowest man and that of the highest animal is immense' (1871:104). One hundred and twenty years later, our understanding of the relationship between the 'mental faculties' or 'minds' of humans and of their closest phylogenetic relations is little more consistent than was Darwin's. (Galef 1992)

Imitation is natural to man from childhood, one of his advantages over the lower animals being this, that he is the most imitative creature in the world and learns at first by imitation. (Aristotle 1941, 4486) (Hauser 1996)

By 'culture' we mean the ability to acquire valuable knowledge and skills from other individuals through social learning and teaching, and to build on this reservoir of shared knowledge, iteratively, generation after generation, building ever more efficient solutions to life's challenges (Laland and Hoppitt 2003).

… we criticize the dominant ('ethnographic') method for identifying animal cultures, which seeks to isolate cultural variation by ruling out alternative explanations for behavioural differences, and instead advocate a more thorough consideration of the interplay between genetics, ecology and culture. (Laland and Janik 2006)

Other animals, including fishes, are capable of social learning and traditionality in behaviour, and in many respects these resemble aspects of human culture and cognition. …how could the extraordinary and unique human capacity for culture have evolved out of something resembling the simple behavioural traditions observed in other animals? (Laland et al. 2011)

Our study adds new evidence supporting the hypothesis that some of the behavioral diversity seen in wild chimpanzees is the result of social transmission and can therefore be interpreted as cultural, especially when considered together with previous results from the wild [10] and captivity [48]. (Hobaiter et al. 2014)

J. A. Colombo, *Dominance Behavior*, https://doi.org/10.1007/978-3-030-97401-5_6

The Anthropomorphic Concept of Culture

The concept of culture and its presence on comparative evolutionary grounds have elicited contradictory opinions in sociology and anthropology. In some authors predominate an anthropomorphic view, thus reducing its presence to human and non-human primate species. We adopt an evolutionary standing on this issue that sustains the view anticipated in previous chapters. It considers culture as representing a social expression whose evolution depended on species-specific constructs—and ecological interactions—thus, expressing aliquots of what represents a natural life mosaic of variable, complex, interactive social construction, and output.

In this context, as stated by Ratnieks and Helantera (2009):

> … social evolution can reach diverse outcomes, which in large part arise from the fact that human society is based on cooperation among unrelated individuals while insect societies are based on altruism among family members (Ratnieks 2006; Ratnieks and Wenseleers 2008).

According to these authors, while inequality and social harmony would appear incompatible or unstable as a social construct from a human perspective, some insect societies such as the honeybee *Apis mellifera* are supremely unequal and harmonious. In these cases, the concept of freedom is severely downgraded by strict censorship control—as it also takes place in authoritarian political institutions expressed along with human history—and dependent on genetic and early preferential breeding. At this point, it appears necessary to close the gap with human social conditions imposed by social and economic inequality that impinges on the probability of differential cognitive development and social insertion in a highly competitive sociality.

As stated by Whiten et al. (2011):

> … the evolution of social learning, traditions and other culturally related phenomena, which have proved to be far more widespread across the animal kingdom than imagined a half-century ago, and more complex in their manifestations…

And, expanded more recently (Whiten et al., 2021):

> It could be said that all culture embodies collective knowledge insofar as the mentality underlying some specific traditional behaviour is distributed across the population of interest. Recognizing this, but going far beyond it, articles in this issue address the significance of cases where collective knowledge exists only at the level of the collective, and not in any one individual. The partial or variant knowledge different individuals that this implies may be distributed across a population in space, and/or over time, with subsequent combinations thence leading to innovations that can drive CCE [82,93,153–155,159,160].

According to the authors, recent archeological developments have pushed back the origins of human culture, suggesting previously unidentified continuities between animal and human culture (Whiten et al. 2011).

Considering that social behavior evolves under specific combinations of relatedness, benefit, and cost according to a central theorem of inclusive fitness (Hamilton's rule) as stated by Bourke (2014), social learning is a universal behavior and a key ingredient of culture. Animals learn skills representing cultural adaptations that affect population structures and the physical and social environment that elicit genetic evolution (Whitehead et al. 2019). Among those skills is niche selection and construction that are reliably transmitted through behavioral tradition across generations and adapted to changing environmental conditions, thus suggesting some form of cumulative adaptive resources expressed in behavioral strategies. Yet, as stated in Stout and Hecht (2017), although niche construction is not unique to humans, no other species approaches the extent, diversity, and complexity of human culture. Hence comparative richness and adaptability as observed in humans introduce a relative consideration that some authors take as an absolute criterion to define culture, as mentioned earlier. For example, Buckholtz and Marois (2012) raise the intriguing possibility that third-party norm enforcement is uniquely human behavior, not shared by chimpanzees, although they engage in costly second-party punishment.

Hence a sort of evolutive gradient on evolutionary standing draws a variable scaling up of social and environmental interactions on species-specific grounds. Tinted with an evolutionist flavor that recalls previous comments on sociality among non-human species, Boehm (2000) poses an intriguing question that draws on political grounds:

> As a species, are we innately given to ethological egalitarianism, to ethological despotism, or to neither? (Boehm 2000)

As briefly mentioned in the above excerpts, attempts to conceptualize an operative and comparative concept of culture represent a debatable issue in the field of behavioral evolution and behavioral anthropology. The limits imposed by each definition determine its projection on constructing an evolutive concept, some of which are tailored to fit human behavior only. The proposal for a conceptualization based on the dynamics of transmittable elementary behavioral cues follows. These allow for constructing an alternative, adaptive behaviors within a multilevel complexity dictated by species differences in neurobehavioral adaptations and ecological contexts.

A predominant concept of *culture* in anthropologic and sociologic resorts seems based on a series of anthropomorphic conditions. It would attempt to contain the complexity and conceptual richness of human material, its social traditions, and creativeness. However, at the same time, it reduces or closes any reference to the evolutive domain from which it emerged.

From a historical perspective, anthropologist Edward B. Tylor in his 1871 work, *Primitive Culture: Research into the Development of Mythology, Philosophy, Religion, Language, Art, and Customs* (cf. Medeiros and Cowall 2020), defined culture as:

> … that complex whole which includes knowledge, belief, art, law, morals, custom, and any other capabilities and habits acquired by man as a member of society.

According to the American Sociological Association:

> Sociology understands culture as the languages, customs, beliefs, rules, arts, knowledge, and collective identities and memories developed by members of all social groups that make their social environments meaningful.[1]

According to the view of the American Anthropological Association (as published in Medeiros and Cowall 2020):[2]

> Culture: a set of beliefs, practices, and symbols that are learned and shared. Together, they form an all-encompassing, integrated whole that binds people together and shapes their worldview and lifeways.

Hence, human-tailored made definitions disregard—does not leave room for—accumulated evidence of a species-specific concept of culture, obtained from comparative studies as outlined in previous chapters and will be considered in the paragraphs below. These latter studies place the concept of culture within a comparative context that includes *Homo* evolution, primate and non-primate mammals, fish, and invertebrate (insect) species.

In their discussion on the conceptual construction and its relevance across animal species of the term *culture,* Laland and Hoppitt (2003) propose the following definition:

> Cultures are those group-typical behavior patterns shared by members of a community that rely on socially learned and transmitted information.

These authors limit its application to a few species besides humans. However, as it will be considered later, recent observations tend to spread the application of culture to several species, including insects. The following authors provide a more plastic and comprehensive evolutionary perspective.

According to Whiten (2005):

> This leads us to define culture as the possession of multiple traditions, spanning different domains of behaviour, such as foraging techniques and social customs.

Further adding:

> As observations accumulated in the 1980s and 1990s, researchers began to compile a growing list of putative traditional variations among wild chimpanzees[2,18]. More recently.... study sites have pooled their hardearned data to create the first systematic overview of behavioural variation. Identifying no fewer than 39 different traditions[12,13].

And in orangutans:

> ... subsequently identified [19] clearly defined traditions, with an additional five more tentatively classified[14].

As Whitehead (2017) states:

> Culture, as an inheritance system, can be defined as behavior or information shared within a community that is acquired from conspecifics through some form of social learning.

---

[1] https://www.asanet.org/topics/culture

[2] https://perspectives.pressbooks.com/chapter/the-culture-concept/

Hence, to attempt defining an evolutive and comparative inclusive concept of culture that would allow for a consideration of other—species-specific—levels of individual and collective (social) behavioral profiles, the following proposal of culture is based on the previous concept:

> *Culture is here understood as a shared, species/order-specific, collective set of physical and social cues and skills transmittable by social learning and imitation that modify individual behavior and imply possessing a socially shared plastic adaptive value.*

Social Transmission of Culture: A Universal Tune Played at Different Scales in the Animal Kingdom

According to Fragaszy and Perry (2003) (cf. Whiten and Schaik 2007), tradition would represent:

> … a distinctive behaviour pattern shared by two or more individuals in a social unit, which persists over time and that new practitioners acquire in part through socially aided learning.

This is a definition that could well represent the concept of culture, according to Laland and Hoppitt (2003). The discussion in Whiten and Schaik (2007) reveals the quagmire it represents for the social sciences, as stated by Laland and Hoppitt (2003) and the debatable inclusive criteria. As mentioned by Whiten and Schaik (2007), culture would represent:

> … the possession of multiple traditions, spanning different domains of behavior.

As the authors' further state:

> The emerging picture of social learning, traditions and culture appears to have the form of a pyramid, the base of which represents *the occurrence of social information transfer of many different kinds in many taxa of the animal kingdom* (Figure 4; see also van Schaik et al. 1999; van Schaik 2004; for related evolutionary models). Social learning, at least in its simpler forms, is well established in mammals, birds and fish (Laland and Hoppitt, 2003) and recent reports extend to invertebrates (Worden and Papaj, 2005). (Italics added by JAC)

The inclusion of several species and orders of the natural kingdom as mentioned earlier and the above statements justifies the concept of culture proposed above, as it includes concepts recognized by Laland and Janik (2006), Laland et al. (2011), Whiten et al. (2011, 2017), and Allen et al. (2013). This allows placing the concept of culture under an evolutive perspective.

In terms of the expression of culture within the context of nonhuman primates, as stated by Whiten et al. (1999) in chimpanzees:

> Among mammalian and avian species, cultural variation has previously been identified only for single behaviour patterns, such as the local dialects of song-birds[12,13]. The extensive, multiple variations now documented for chimpanzees are thus without parallel. Moreover, the combined repertoire of these behavior patterns in each chimpanzee commu-

> nity is itself highly distinctive, a phenomenon characteristic of human cultures[14] but previously unrecognized in non-human species.

As stated by van Schaik et al. (2003) in orangutans:

> … as expected under a cultural interpretation, we find a correlation between geographic distance and cultural difference, a correlation between the abundance of opportunities for social learning and the size of the local cultural repertoire, and no effect of habitat on the content of culture. Hence, great-ape cultures exist, and may have done so for at least 14 million years.
>
> Human cultures, therefore, differ from those of great apes in having unambiguously symbolic elements (6, 27), far more complex skills, and far greater repertoire sizes, made possible by cognitive differences affecting innovation or observational learning (1, 5, 6). The presence in orangutans of humanlike skill (material) culture pushes back its origin in the hominoid lineage to about 14 million years ago, when the orangutan and African ape clades last shared a common ancestor (28), rather than to the last common ancestor of chimpanzees and humans.

As stated by Koops et al. (2014):

> As there is now overwhelming evidence from wild and captive studies that tool use in primates is socially learned [18], the following cases are assumed to be culturally transmitted.

The following selected statements provide further grounds to the conclusion stated above:

> However, although such studies shed light on social transmission processes and the capabilities of particular species, at best, they provide circumstantial support for the argument that natural behaviour is cultural. (Laland and Janik, 2006)

> Recent studies reveal that processes important in cultural transmission are more widespread and significant across the animal kingdom than earlier recognized, with important implications for evolutionary theory… These developments suggest previously unidentified continuities between animal and human culture. (Whiten et al. 2011)

> Other animals, including fishes, are capable of social learning and traditionality in behaviour, and in many respects these resemble aspects of human culture and cognition.
>
> In many respects, the social learning of fishes is very much comparable to that of birds and mammals. Similar experiments in mammals and birds, such as transmission chains or diffusion studies, generate broadly similar patterns, and appear to be reliant on similar processes, as those described in fishes [74–77]. While individual species or genera of primates, such as chimpanzees, orangutans, macaques, and capuchins, may arguably exhibit more sophisticated cultural capabilities. (Laland et al. 2011)

> These insights strengthen the case that cetaceans represent a peak in the evolution of nonhuman culture, independent of the primate lineage. (Allen et al. 2013)

It is worth mentioning that Allen et al. (2013) points to the fact that humpback whale populations carry vocal constructions in the form of songs, hence implying that this population can be considered to carry multiple traditions.

More recently, Whiten et al. (2017) (see also in Whiten et al., 2021) advanced conclusive statements regarding the diffusion of cultural strategies among orders and species:

> Social learning is now extensively documented in mammals (29), with a particular intensity of research studies in primates (30–33) and cetaceans (34–37), in birds (38–41), in fish (42), and in insects (43, 44). The fact that social learning has been shown to play important roles spanning multiple functional contexts (25–28) suggests that many animals are not simply acquiring one or a few behavioral patterns socially, but rather that social learning is central to their acquisition of adaptive behavior.
>
> …cultural diffusion has been further documented not only in all of the vertebrate research reviews listed above, but also in insects, at least in the laboratory.
>
> Recent evidence that insects also show not only social learning, but a capacity for cultural transmission spreading across communities.

It further states that:

> However, there is evidence that animal traditions with suboptimal payoffs are sometimes, although seemingly not always (85), vulnerable to decay (41) … Nonetheless, experimental studies, for example, of mate choice copying, show that animal social transmission can be evolutionarily consequential, even if short lived (86).

This provides the basis for an evolutionary perspective regarding animal cultural evolution based on universal constants—besides species-specific interactive domains—as anticipated in the following statement (Whiten and Mesoudi, 2008), later supported by Alem et al. (2016) and Leadbeater and Dawson (2017) observations in insect species:

> … we summarized the current picture for fishes, birds, rodents and primates as it currently appears, suggesting both that elementary forms of cultural transmission are widespread across this taxonomic range, and that more complex contents and mechanisms are identifiable in the avian and mammal studies, particularly in the primate ones where they extend to tool use and more elaborate manipulative and foraging techniques. This pattern suggests a series of phases in the evolutionary elaboration of cultural transmission that paved the way for human culture. (Whiten and Mesoudi 2008)

Recent evidence in insects also shows not only social learning (see in Worden and Papaj, 2005) but a capacity for cultural transmission spreading across communities, as studied by Alem et al. (2016) and by Leadbeater and Dawson (2017) in bumblebees and honeybees. Interestingly, Worden and Papaj (2005) further include basic statements regarding that bumblebees show a novel form of social learning in insects of unique benefits for the colony and that:

> …evidence continues to mount that there is no strict dichotomy between vertebrate and invertebrate cognition (Fiorito and Scotto 1992; Giurfa et al. 2001; Srinivasan and Zhang 2003) and that brain size does not necessarily limit an animal's (*basic*) cognitive abilities. (Italics added by JAC)

In this regard, Leadbeater and Dawson (2017) state that:

> Social insects are distant relatives of vertebrate social learners, but the research we describe highlights routes by which natural selection could coopt similar cognitive raw material within the animal kingdom.

Alem et al. (2016) succeeded to demonstrate in bumblebees that learning a nonnatural task can spread culturally through populations:

> Our results indicate that bees may not be different from birds, dogs, or apes in this respect. Bumblebees relied on the perceptual feedback provided by their actions…

The authors provide further insight into evolutionary precursors of human cultural sophistication:

> Our findings add to the accumulating evidence suggesting that the capacity of culture may be within most animals with a relatively basic toolkit of learning processes as described here, in turn shedding light on the evolutionary precursors of the more sophisticated forms of culture in humans.

It further adds that:

> … This suggests that, so long as animals have a basic toolkit of associative and motor learning processes, the key ingredients for the cultural spread of unusual skills are already in place and do not require sophisticated cognition.

Chittka (2017) and Loukola et al. (2017) support the concept that small insect brains can provide basic components and processing involved in cognition, though using paradigms removed from problems encountered naturally (Loukola et al. 2017). This concept implies that bees' brains are wired for cognition, extracting rules from the environment, predicting future alternative strategies based on efficient information storage and retrieval, even when submitted to a novel, experimental environment.

As Leadbeater and Dawson (2017) clearly state:

> An expanding body of work now shows that social learning (1), once considered the preserve of vertebrate species, is a feature of insect behavioral repertoires (2, 3). Insects not only learn about foraging skills, food preferences, brood hosts, and potential mates by responding to information provided inadvertently by others (4–11), but also transmit these behaviors further, such that they propagate through groups (8, 12) and possibly even through wild populations (13). Some of these phenomena appear similar to socially learned behavior patterns that have been described in vertebrates (11, 14) at least outside the context of imitation, and as such they are interesting extensions to the taxonomic distribution of social learning.

Interestingly the authors state that information provided inadvertently by other foraging bees influences this learning process, a simple example whereby social bees learn from their conspecifics about rewarding flower types, implying an instance of social learning. As the authors conclude, social learning abilities may have traveled further along evolutionary routes in some but not in all lineages.

Slagsvold and Wiebe (2011) provide evidence of learning transmission in birds, suggesting that foraging behavior can be culturally transmitted over generations in the wild and may have ecological and evolutionary consequences. According to Rendell et al. (2011), this behavior would be like that observed in fish. Comparison with subtleties and complexity of human culture should not preclude sharing basic strategies with other species, as mentioned by Rendell et al. (2011):

> An important contribution of this work, in parallel with studies on non-humans, is to challenge the notion of a single best strategy ***(for social learning)***, or a strategy associated with a particular type of individual, or species. (Bold characters inserted by JAC)

In many respects, the social learning of fishes is very much comparable to that of birds and mammals. Similar experiments in mammals and birds, such as transmission chains or diffusion studies, generate broadly similar patterns and appear to be reliant on similar processes, as those described in fishes [74–77]. While individual species or genera of primates, such as chimpanzees, orangutans, macaques, and capuchins, may arguably exhibit more sophisticated cultural capabilities.

The specific characteristics expressed in human culture involving cooperative breeding and allomaternal care—universals of human culture—are considered to have distinct evolutionary origins. These are stressed by Burkart et al. (2017), aside from critical mechanisms shared across species, suggesting a continuing—not necessarily linear—evolutionary path.

Our Species *Subjacent Ancestral nature*

Though the emergence of our species cultural profile remains unsettled, as mentioned in d'Errico and Stringer (2011):

> Three scenarios have been proposed to account for the origin of cultural modernity. The first argues that modern cognition is unique to our species and the consequence of a genetic mutation that took place approximately 50 ka in Africa among already evolved anatomically modern humans. The second posits that cultural modernity emerged gradually in Africa starting at least 200 ka in concert with the origin of our species on that continent. The third states that innovations indicative of modern cognition are not restricted to our species and appear and disappear in Africa and Eurasia between 200 and 40 ka before becoming fully consolidated.

Territoriality and prevalence are universal domains in survival behaviors within the natural kingdom. Whether having solitary or gregarious social habits, animals and vegetables (Mancuso 2019;[3] Baluska and Mancuso 2020) show territorial behaviors toward conspecifics and pray on them or reject those that compete with feeding or reproductive resources.

Our species biocultural origin has its roots in ancestral habits, behaviors, and survival drive, through changing environmental conditions, and crystallized during millennia in basic neurobehavioral circuits, be it as predators or potential prey. Besides new sets of genes that would condition new brain developments, *Homo sapiens* carries an essential, ancestral backpack imprinted in basal brain neural circuits subserving basic needs and drives. Its culturally transformed—or hidden—expression occurs as dominance/prevalence behavior, thrive for survival, and hierarchical social constructs, either spontaneously, under dynamic social circumstances, or as transformed into virtual (cultural, ideological) domains. Its relationship with the continuous development of sophisticated material culture interactively

[3] https://www.ted.com/talks/stefano_mancuso_the_roots_of_plant_intelligence?language=es

evolves our collective mind and virtual constructions. *Homo sapiens* evolution depends on the construction of instruments of progressive complexity and power utility, developed into a—material—cultural technology that resets the relationships among individuals and between them and the environment.

Sociocultural differences among worldwide communities—and within them, also among its constituents—result from a different history and dynamics of genomic and environmental interactive human constructions. This concept supports limiting the current concept of globalization to limited strata of the socioeconomic domain.

Which, and how much, of our current behaviors—individually and as a global community—are driven by ancestral, inherited traits imprinted in our animal condition? The cultural domain implied in this question pertains to our identity and pertinence to social constructions and ecological interaction. Yet all are not events of the conscious dimension. Cognitive processing involves distributed neural circuits as a substrate. Perhaps the most disturbing from an intellectual point of view is that much of the former appears to be at the subconscious level. According to some authors, what emerges at the conscious level, expressed temporarily at a specific time, are the events we can manipulate as working memory in our executive behavior domain, corresponding to the explicit memory (Dietrich 2015).

The *ancestral animal nature* (animal drives), progressively built and placed into practice since primeval times, is crimped to our phylogenetically old, basic neural systems and behavioral survival construction. It keeps playing a role in our social interactions and cultural construction (Colombo 2019, 2020a, b) and expressed in response to multiple challenges. No wonder our species has been considered to have a bipolar behavioral profile (Boehm 2012a, b) attributable to genetic lineages spanning from the *ancestral Pan*, profiled by the behaviors of chimpanzees (tendency to conflict, male predominance) and bonobos (preventive behavior, female predominance), this, besides the cultural environment and set of values that each ethnos has interactively developed for itself. It anticipates a potential behavioral bipolarity with uneven prevalence distribution among individuals and social organizations.

Through time, based on social repression or socialization, cultural strata of variable thickness have been constructed on top of drives implicit to our animal condition. Nevertheless, as the history of human civilization demonstrates, it failed in their deactivation or suppression and only succeeded in reformulating or transitorily repressing them and generating socially deviant behaviors expressed at the individual and collective levels. Interaction with the physical and cultural environment continues to remodel our ethnic profiles. However, our primary organization is bound to ancestral demands that imprinted a given set of basic drives (territorialism, reproductive, survival, secure feeding sources, dominance/prevalence, accumulative behavior). Their expression, affected by changed environmental conditions (physical and sociocultural interactions), poses the probability of continuous frictions between the *neurobiological* and *cultural tectonic plates*, as illustrated in Fig. 11.1 (Colombo 2010, 2015).

Indeed, our brain plasticity and mind construction (depending on cultural issues and contexts) provide for adaptative responses. However, so far, these have not canceled the framework of subjacent, primary drives (as mentioned above) imprinted in the heart of our neurobehavioral animal construction, although social (cultural) settings may affect the probability or profile of their expression. That is, biosocial interactions continue to model phenotypes and behavioral trends on top of the basic, deeply entrenched, survival, and prevalent drives, conforming to the basic structure of the ancestral animal nature.

Among humans, this has conditioned the disparate social, cultural, and cognitive conditions among nations, ethnics, and individuals that have contributed to building the *crackled* composition of our modern world. It applies to distribution of economic resources, political and financial dominance/prevalence, unequal rights, poverty, cultural profiles, and disparate unequal financial wealth, as well as to relative access to cognitive development and quality of life. The inequality of its distribution fuels the continued state of instability among nations and within them. In this political domain, as Hill (1993) states in a commentary:

> An implicit corollary to Boehm's hypothesis is that leaders organize followers' intentions to construct hierarchy on a consensual basis.

References

Alem S, Perry CJ, Zhu X, Loukola OJ, Ingraham T, Søvik E, Chittka L. Associative mechanisms allow for social learning and cultural transmission of string pulling in an insect. PLoS Biol. 2016;14(12):e1002589. https://doi.org/10.1371/journal.pbio.1002589.

Allen J, Weinrich M, Hoppitt W, Rendell L. Network-based diffusion analysis reveals cultural transmission of lobtail feeding in humpback whales. Science. 2013;340:485–8.

Baluška F, Mancuso S. Plants, climate, and humans. Plant intelligence changes everything. EMBO Rep. 2020;21(3):e50109. https://doi.org/10.15252/embr.202050109.

Boehm C. Conflict and the evolution of social control. J Conscious Stud. 2000;7:79–101.

Boehm C. Costs and benefits in hunter-gatherer punishment. Behav Brain Sci. 2012a;35:19–20. PMID: 22289308

Boehm C. Ancestral hierarchy and conflict. Science. 2012b;336:844–7. https://doi.org/10.1126/science.1219961.

Bourke AFG. Hamilton's rule and the causes of social evolution. Philos Trans R Soc Lond Ser B Biol Sci. 2014;369(1642):2013.0362. https://doi.org/10.1098/rstb.2013.0362.

Buckholtz JW, Marois R. The roots of modern justice: cognitive and neural foundations of social norms and their enforcement. Nat Neurosci. 2012;15:655–61.

Burkart JM, Schubiger MN, van Schaik CP. The evolution of general intelligence. Behav BrainSci. 2017;40:e195. https://doi.org/10.1017/S0140525X16000959.

Chittka L. Bee cognition. Curr Biol. 2017;27:R1037–59.

Colombo JA. Somos la especie equivocada? (Are we the wrong species?). Buenos Aires: Ed. EUDEBA; 2010.

Colombo JA. Los Homo sabios (The wise Homo). Buenos Aires: Ed. Buenos Aires Books; 2015.

Colombo JA. Our animal condition and social construction. New York: Nova Science Publishers; 2019. ISBN: 978-1-53615-357-6. eBook ISBN: 978-1-53615-3583

Colombo JA. Creativity, a profile for or species. Newcastle: Cambridge Scholars Publications; 2020a.
Colombo JA. Animal heritage vectors and human social construction. 3rd International conference on advanced research in social sciences and humanities. 2020b.
D'Errico F, Stringer CB. Evolution, revolution, or saltation scenario for the emergence of modern cultures? Philos Trans R Soc B. 2011;366:1060–9.
Dietrich A. How creativity happens in the brain. New York: Palgrave-Macmillan; 2015.
Galef BG Jr. The question of animal culture. Hum Nat. 1992;3:157178.
Hauser MD. The evolution of communication. Cambridge: MIT Press; 1996.
Hill JD. Commentary. Curr Anthropol. 1993;34:242–3.
Hobaiter C, Poisot T, Zuberbuhler K, Hoppitt W, Gruber T. Social network analysis shows direct evidence for social transmission of tool use in wild chimpanzees. PLoS Biol. 2014;12(9):e1001960. https://doi.org/10.1371/journal.pbio.1001960.
Koops K, Visalberghi E, van Schaik CP. The ecology of primate material culture. Biol Lett. 2014;10:20140508. https://doi.org/10.1098/rsbl.2014.0508.
Laland KN, Hoppitt W. Do animals have culture? Evol Anthropol. 2003;12:150–9. https://doi.org/10.1002/evan.10111.
Laland KN, Janik VM. The animal cultures debate. Trends Ecol Evol. 2006;21:542–7.
Laland KN, Atton N, Webster MM. From fish to fashion: experimental and theoretical insights into the evolution of culture. Philos Trans R Soc B. 2011;366:958–68. https://doi.org/10.1098/rstb.2010.0328.
Leadbeater E, Dawson EH. A social insect perspective on the evolution of social learning mechanisms. PNAS. 2017;114:7838–45.
Loukola OJ, Perry CJ, Coscos L, Chittka L. Bumblebees show cognitive flexibility by improving on an observed complex behavior. Science. 2017;355:833–6. https://doi.org/10.1126/science.aag2360.
Medeiros P, Cowall E. The culture concept. In: Brown N, McIlwraith T, Tubelle de González L, editors. Perspectives: An open introduction to cultural Anthropology. 2nd ed. Arlington: The American Anthropological Association; 2020. https://perspectives.pressbooks.com/chapter/the-culture-concept/.
Ratnieks FLW, Helantera H. The evolution of extreme altruism and inequality in insect societies. Philos Trans R Soc B. 2009;364:3169–79. https://doi.org/10.1098/rstb.2009.0129.
Rendell L, Fogarty L, Hoppitt WJE, Morgan TJH, Webster MM, Laland KN. Cognitive culture: theoretical and empirical insights into social learning strategies. Trends Cogn Sci. 2011;15:68–76. https://doi.org/10.1016/j.tics.2010.12.002.
Slagsvold T, Wiebe KL. Social learning in birds and its role in shaping a foraging niche. Philos Trans R Soc B. 2011;366:969–77. https://doi.org/10.1098/rstb.2010.0343.
Stout D, Hecht EE. Evolutionary neuroscience of cumulative culture. PNAS. 2017;114:7861–8. https://doi.org/10.1073/pnas.1620738114.
van Schaik CP, Ancrenaz M, Borgen G, Galdikas B, Knott CD, Singleton I, Suzuki A, Utami SS, Merrill M. Orangutan cultures and the evolution of material culture. Science. 2003;299:102105.
Whitehead H. Gene -culture coevolution in whales and dolphins. PNAS. 2017;114:7814–21. www.pnas.org/cgi/doi/10.1073/pnas.1620736114
Whitehead H, Laland KL, Rendell L, Thorogood R, Whiten A. The reach of gene-culture coevolution in animals. Nat Commun. 2019;10:2405. https://doi.org/10.1038/s41467-019-10293-y.
Whiten A. The second inheritance system of chimpanzees and humans. Nature. 2005;437:52–5.
Whiten A, Mesoudi A. Establishing an experimental science of culture: animal social diffusion experiments. Philos Trans R Soc B. 2008;363:3477–88. https://doi.org/10.1098/rstb.2008.0134.
Whiten A, van Schaik CP. The evolution of animal 'cultures' and social intelligence. Philos Trans R Soc B Biol Sci. 2007;362:603–20. https://doi.org/10.1098/rstb.2006.1998.
Whiten A, Goodall J, McGrew WC, Nishida T, Reynolds V, Sugiyama Y, Tutin CEG, Wrangham RW, Boesch C. Cultures in chimpanzees. Nature. 1999;399:682–5.

Whiten A, Hinde RA, Laland KL, Stringer CB. Culture evolves. Philos Trans R Soc B. 2011;366:938–48. https://doi.org/10.1098/rstb.2010.0372.

Whiten A, Ayala FJ, Feldman MW, Laland KL. The extension of biology through culture. PNAS. 2017;114:7775–81. www.pnas.org/cgi/doi/10.1073/pnas.1707630114

Whiten A, Biro D, Bredeche N, Garland EC, Kirby S. The emergence of collective knowledge and cumulative culture in animals, humans and machines. PhilosTrans R Soc B. 2021;377:20200306. https://doi.org/10.1098/rstb.2020.0306.

Worden BD, Papaj DR. Flower choice copying in bumblebees. Biol Lett. 2005;1:504–7. https://doi.org/10.1098/rsbl.2005.0368.

Chapter 7
Means of Social Coercion

> Many aspects of our human condition evolved in the deep past and cannot be fully understood without the long vantage point of history and prehistory. This is certainly true for the fundamental principles of human organization –the structure and function of the operation of society– which have been present for thousands of years.
>
> A fair question might be raised at this point –why are we still talking about the emergence of social inequality? The simple answer is that we still don't know very much. The origin of inequality remains essential because there is no scholarly consensus… We do know that social inequality has been the dominant structuring principle in most human societies over the last 5,000 years or more. At the same time, we still do not know precisely when or why this principle became dominant, or how it operated in the past. (Price and Feinman 2010)

> … cultural behaviour may select for particular functional genes, influence patterns of genetic diversity, and spark speciation. When cultural activity is an important determinant of fitness, it can generate selection for traits that further enhance cultural competencies, allowing genes and culture to coevolve reciprocally. (Whitehead et al. 2019)

As stated by Lahr and Foley (1994), Templeton (2002) (cf. Etler, 2005), and Beyin (2006, 2011), recent human (*Homo sapiens*) evolution was influenced by two significant events that encompass at least two major expansions during the upper Pleistocene colonization of Eurasia—through Ethiopia/Arabia toward South Asia and through North Africa/Middle East toward Eurasia. This took place following the range extension of *Homo erectus* out of Africa, with genetic exchange among migrated *Homo* populations at different times. The latter also involved potential adaptations to diverse ecological settings through millennia. Thus, multilateral dispersion of *Homo* and more recent *H. sapiens* populations built the probability of genetic exchange and behavioral profile encounter with genotypic expression and phenotypic adaptations.

An additional comment may complete the picture of our complex origin as a species. As quoted by Scally et al. (2012):

> Since the middle Miocene—an epoch of abundance and diversity for apes throughout Eurasia and Africa—the prevailing pattern of ape evolution has been one of fragmentation and extinction[48]. The present-day distribution of non-human great apes, existing only as

J. A. Colombo, *Dominance Behavior*, https://doi.org/10.1007/978-3-030-97401-5_7

endangered and subdivided populations in equatorial forest refugia[43], is a legacy of that process. Even humans, now spread around the world and occupying habitats previously inaccessible to any primate, bear the genetic legacy of past population crises. All other branches of the genus *Homo* have passed into extinction. It may be that in the condition of *Gorilla, Pan* and *Pongo* we see some echo of our own ancestors before the last 100,000 years, and perhaps a condition experienced many times over several million years of evolution. It is notable that species within at least three of these genera continued to exchange genetic material long after separation.[4,49]

Thus, besides the species-specific phenotypic complexities, cooperation and competition would express universal components of the natural kingdom. These evolved in *Homo* under the pressure of diverse genetic and phenotypic encounters and interactions that progressively resulted in present-day human diversity. Cooperation and competition have molded animal groups' social configuration and survival probability, from insects to mammals. These two semantically opposing behavioral components represent a conundrum for human evolution, in which the complexity of behavioral drives exceeds those of reproduction and feeding. Although, as stated by Nowak (2006):

> From hunter-gatherer societies to nation-states, cooperation is the decisive organizing principle of human society. No other life form on Earth is engaged in the same complex games of cooperation and defection.

It should be underlined that engaging in a competition aimed at gaining social status has been a universal component in the natural kingdom, including among and within human cultures. Vested interests have been a constant, which has approached dangerous limits for survival and acceptable life quality. Among humans, political and corporative dominance/prevalence has acquired sophisticated means to impose social values, goals, and supposed material needs, using diverse overt and subliminal strategies.

Moral Values and Religiosity as Social Controllers

> Morality today involves social control but also the management of conflicts within the group. It is hypothesized that early manifestations of morality involved the identification and collective suppression of behaviours likely to cause such conflict. (Nowak 2006)

The following excerpt, referred to Finland's prehistoric history, provides an example of a universal cognitive domain:

> Belief systems evolved organically as part of social adaptation strategies. They conform to a specific dialogue between man and habitat and represent an intrinsic element of daily life. They were integrated to the logic of subsistence. (Colombo and Kokkonen 2018)

Communities thrive based on shared values that define the range of acceptable behaviors, reinforced by systems of sanctions and punishments of deviant behaviors. These values are expressed on the layman system of rules and in religious commandments—sanctions in each case aimed at either physical, social, or emotional dimensions. While the formers are based on debatable principles for those that infringe the law, the latter recalls emotional fears on virtual agencies on which religiosity stands. Each usually operates on some form of coalitions or grouping, thus finding further support for their standings. Besides the formal complexities that social sanctions might express, its origin recognizes comparative evolutionary antecedents across species. In other words, social coercion and coalitions find common expression across natural history, based on hierarchical or gregarious contexts. Such is the case of social castes described in insects, pecking order among birds, and sexual dimorphic changes among group dynamics in the coral fish, or in mammalian species (herds, packs) and primates, institutionalized autocratic systems are expressed among humans. In every case, social standing is based on power control, dominance/prevalence, and submission, or acceptance, of basic rules. The rate of evolution of social structures in nonhuman primate and human societies—in terms of individual and group development—is based on a required degree of interaction and social mobility, that is, on a given degree of entropy within the system. Hence, differences between social mobility drives and predatory behaviors enter the list of components—a list that is often established or modulated by dominant groups—to be distinguished and allowed or sanctioned. It seems appropriate to include here the exception implied in egalitarian societies, as stated in Boehm (2000):

> However, an older type of human society is the politically-equalized group I have described above. Vehrencamp (1983) would designate this ethologically as an egalitarian society, for the expression of tendencies to dominance and submission is not pronounced. This is because individual rank is not a major factor in reproductive success, and because leaders are not very dominant.

Based on genotypic and phenotypic characteristics, social species in the natural kingdom regulate individual interactions according to norms enforced by direct (physical) and indirect (feeding availability, reproductive allowances, cultural rules) coercion. These represent forms of dominance/prevalence expressed economically, ideologically, or through institutionalized contracts among humans. So, collective values recognize an evolutionary expression of ancestral, basic, prosocial normative, or moral notions. Regarding cooperation, according to Tomasello (2018):

> … human morality comprises the psychological dimensions of humans' species-unique forms of cooperation.

This behavior would not be exclusive to humans since, as stated by Warneken and Tomasello (2006) (cf. Tomasello, 2018), chimpanzees will help others attain their goals if the situation is such that there are no self-interested or competitive motives in their way. It implies shared intentionality, also shown by hyenas, lions, and chimpanzees during group hunting or territorial defense. Human culture underlines the complex weaving of interactions in various behavioral domains, as mentioned by Schuster (2005):

> Shared intentionality may be more obvious in humans because its widespread use –in culture-based activities such as science, art, music, and religion– was facilitated by language and culture by a process of exaptation.[1]

[1] Process by which features acquire functions for which they were not originally adapted or selected.

Interactions between individual and collective interests are shaped according to species-specific social construction, i.e., expression of cooperative and competitive behaviors is molded according to genetic, relational (kin), and phenotypic configuration. These conditional interactive socio-behavioral patterns have been analyzed, among others, by Tomasello et al. (2005), Warneken et al. (2007), Schmelz et al. (2017), and Tomasello (2018).

On human developmental grounds, according to Tomasello (2018), at the age of 3 years, a transition would take place in terms of prosociality and moral beings:

> My argument here is simply that whatever the methods or paradigm, it is an empirical fact that at around 3 years of age, young children make a fundamental transition from prosocial beings to moral beings.

Social Inequality and Coercion in Human Evolution

As mentioned by Price and Feinman (2010):

> The origin of inequality remains essential because there is no scholarly consensus…

The central issue of inequality presents at least two intimately related domains. These are within the general Darwinian evolutive and the human sociocultural domains. In each case, dominance/prevalence and survival represent two main ancestral drives that persist in our ultra-social species. The dynamics implicated in human behavior will be considered an evolutive case of universal animal drives. This viewpoint justifies its consideration within the dynamics of the ancestral anlage of our animal origin. That could be why our "rational" species could not "shake off" ancient animal drives that anchor our individual and collective behavior to them, also—within the dominant-submissive-reactive social context—why our "rational" species could not "shake off" the rampant inequality in our human sociocultural systems. In the animal kingdom, inequality represents a universal domain. At the human level, it arises when social rights, access to education, and living condition characteristics are not met or violated; set aside the behavioral deviation that results in diverse forms of physical and emotional forms of torture, a human behavioral acquisition (Colombo 2021a).

As mentioned by Segretin et al. (2014):

> … associations between socioeconomic status and cognitive development during childhood are mediated by biological, psychological and environmental factors, which may be conceptualized at multiple levels of analysis (individual, family, and social contexts), and increase the likelihood of negative impacts later in life (Leinonen et al. 2002; Raver et al. 2007, 2013; Santos et al. 2008; Cadima et al. 2010; Rhoades et al. 2011; Sarsour et al. 2011; Lipina et al. 2013).

According to Whitehead et al. (2019), culture typically leads to the production and propagation of adaptive behavior. Learned behavioral innovation is now extensively documented in several orders and species of the animal kingdom, including insects.

This process results in individuals often copying successful individuals and high-payoff behavior that enhances the spread of adaptive variants. According to these authors:

> Culturally mediated speciation was a topic first introduced for birds[44] and whales[77], but might also apply to hominins. It is now well-established that Neanderthals, Denisovans and Homo floresiensis coexisted with modern humans and possessed distinct behavioural repertoires[116]. Culturally transmitted ecological specializations may have led to radiations comparable to that of killer whales in our own evolutionary past, possibly on multiple occasions.

Culture develops (see Box 7.1) as a process that implies the interaction of creativeness, sharing, institutionalized (or hierarchical) permissiveness, and coercion. An unsolved central issue is its role as a trigger that generates inequality imposed by different means and individual and collective historical trajectories. These conditions impact the brain and mental development. On a comparative and evolutive basis, hierarchical inequality is a standing component in sociality, from insects to primates. According to Cooper and West (2018), division of labor is a driver, not a consequence, of major evolutionary transitions to higher levels of stratified individuality, such as in eusociality. However, we have seen that eusociality involves high degrees of extreme social inequalities that provided means for enduring

Box 7.1: Our Animal Heritage

"Based on the behaviors of the chimpanzee and the bonobo, who share a common ancestor to Homo, Boehm (2012) poses that human would keep behavioral styles from chimpanzees (tendency to conflict, male predominance), and bonobos (preventive behavior, female predominance). This anticipates a potential behavioral bipolarity with uneven prevalence distribution among individuals and social organizations. Based on social repression or 'socialization', cultural strata of variable 'thickness' have been constructed on top of drives implicit to our animal condition. Nevertheless, it failed in their deactivation, only in reformulating or repressing them. Interaction with the physical and cultural environment continues modeling our ethnic variations, yet our primary organization is bound to ancestral demands that imprinted a given set of basic behaviors, as mentioned before (territorialism, sexual drive, individual survival, secure feeding sources). Their expression, affected by changed environmental (physical and social) conditions, pose the probability of frictions between the neurobiological and cultural tectonic plates (Colombo 2010, 2013, 2016). It is true that the surprising plasticity of our brain and mind construction (depending on cultural issues) provide room for adaptative responses. However, they do not cancel the framework of primary drives imprinted in the heart of our animal construction, although they may affect their expression. Psychological mechanisms disguise, or temper, some of the behavioral consequences of such frictions or uncoupling." (Colombo 2021b) success, as observed among insect species. This hierarchical social structure persists among mammalian species with the addition that hierarchies are settled following physical or political competitive instances. Among insects it involves preferential feeding at larval stages. It should be stressed that human feeding and health inequality also have early nutritional consequences on child development and social adaptation in vast sectors of the world population; in this regard, inequality severely affects social, physical, and cognitive development. Human's inadequate nutrition and access to cognitive enrichment expressed in significant population groups represent a spin-off consequence of socioeconomic inequality; the formers were considered counter-evolutionary conditions (Colombo 2010).

According to Mark Hereward (UNICEF):[2]

> In 2019, 33 million children were displaced by violence or conflict. This includes 19 million children internally displaced within their own countries, 13 million refugee children and around 1 million children seeking asylum. Millions more are on the move as seasonal migrants or internal migrants and each day more are displaced by climate-induced disasters. Uprooted from their lives, these children are among the most vulnerable in the world. They have limited access to education, health care, clean water or protective services and the COVID-19 pandemic has only further eroded their precarious situation.

As previously mentioned, the described inadequate—socially degrading—developmental condition, enhanced by its chronic exposure, hampers social and cognitive development. These conditions result in further social stratification and inequality, among other community impacts.

Among theoretical and experimental strategies providing an understanding of biological mechanisms in complex species, there is a rich precedent based on the predictions obtained from simpler organisms. However, the neurobehavioral complexity in higher vertebrates involves additional genetic and intervening environmental factors (social, cultural, and ecological) to the comparatively simpler equation drawn based on less complex organisms, differential neurocognitive connectivity, and domain interactions.

Therefore, the central question is why *Homo sapiens* has not been able to generate conditions to reach a stable egalitarian social mode in terms of institutionalized rights (basically, access to education and early cognitive enhancement and collective early balanced nutrition), unless ancient animal drives still drive us in terms of social construction based on power hierarchy, resulting in privileges and social stratification (Colombo 2021b). As stated by Hayden and Villeneuve (2010):

The ultimate motivation for developing and maintaining status inequality is the benefit conferred on those in power.

One common interactive component among social organisms is coercion exerted by environmental factors, social (or institutional) structures, or individual (physical, emotional) interactions. As was mentioned by Henrich and Boyd (2008), social and

[2] https://data.unicef.org/resources/migration-and-displacement-country-profiles-mdcp/?utm_source=newsletter&utm_medium=email&utm_campaign=migration%20displacement%20country%20profiles

interindividual coercion exists even in primitive human societies, almost probably also long before in most social or relational forms in the natural kingdom.

Among humans, according to Ratnieks and Helantera (2009):

> Advanced societies, such as the modern-day nation states, typically have elaborate ways of coercing group members including institutionalized police forces, taxation, punishment and surveillance. Opting out (except by migrating to another nation, which probably has similar restrictions) is not permitted. Coercion is well known for enforcing inequality, such as when it serves to promote the advantage of an elite.

Humans apply coercion using several strategies, such as social, animistic beliefs, institutionalized mandate, religious beliefs and sense of guilt, group identification, and emotional relatedness. Coercion to abide by given rules and social-behavioral profiles forms part of the socialization and survival process. Prolonged beyond developmental years, it aims at social integration, predominance, or submissiveness. This process is present in different degrees in non-egalitarian species, expressed either as intragroup hierarchies or territoriality, and acquires greater complexity in human societies.

As stated by Price and Feinman (2010):

> At some point in the deep human past, the biological imperative for dominance behavior, common in our closest animal relatives, was dampened by a cultural mechanism. This mechanism, known as egalitarianism, reflects the importance of cooperative behavior in the emergence of culture, in learning and sharing knowledge, and in survival (e.g., Boehm 1993, 2000, Knauft 1991, Wiessner 2002). Human society operates within this didactic tension between dominance and equality, between hierarchical and egalitarian, **between modes of behavior that feature or privilege the group to those that accent individuals**. (Bold characters inserted by JAC)

As described in previous chapters, communal living implies varying degrees of coercion and power delegation. The structure of unequal relations, of status differentiation, is essential to so-called eusocial organizations in which hierarchies develop through genetic and epigenetic processes and competitive strategies for dominance/prevalence. Price and Feinman (2010) stated that the evolution of larger group sizes and the probability of frequent interpersonal interactions interacted with new social arrangements. Boyd and Richerson (1988) support the concept that reciprocal cooperation becomes more difficult with increasing group size. In such circumstances, reciprocal cooperation usually persists as an adaptation and survival resource in economically lower social classes.

According to Boehm (1993), a substantial cross-cultural survey revealed mechanisms by which the political rank created a reverse dominance hierarchy, i.e., concerning the opposed condition as represented by an egalitarian behavior, an instance of followers dominating their leaders. Boehm attributes this reversed hierarchy important implications for cross-phylogenetic comparisons and the theory of state formation.[3] It remains anthropologically unsettled whether in simpler societies it is present the innate notion of social hierarchy. Closely related to this issue is that for

[3] According to Boehm (1993), the central idea of an egalitarian society is that "in such societies political leadership is weak and ranking and stratification among adult males are absent or muted".

some time, the small group size would have made possible an egalitarian structure favored by migration when the numbers tended to alter such conditions. This statement implies that social hierarchies flourished with conditions that favored settlements of larger populations.

Social means for acquiring dominance/prevalence hierarchy is probably entrenched with ancient drives of animal evolution and cumulative culture as another source of power or predominance. As succinctly described by Stout and Hecht (2017) and Stout (2018), the successive early periods of instrumental and technological development in human history are linked with capacities implicit in brain evolution and on evolving brain-behavior-culture interactions.

In modern times, the current socioeconomic trend irrupted in this series of interactive bargain events and the development of a mounting power structure by the advent of the so-called *Homo economicus*, which supposedly represents a financial term that economists use to describe as a rational human being. According to Wilson (2008),[4] this was first proposed by John Stuart Mills in an 1836 essay defining the characteristics of political economy, which would imply the ability to make rational decisions. This contradicts the position of modern behavioral economists and neuroeconomics, which have demonstrated that human beings make mostly predictable irrational decisions, i.e., not devoid of an emotional component. The basic concept of *Homo economicus* and the idea that humans always act rationally are challenged by risk aversion. Interestingly, before *Homo economicus*, Rochat (2005) proposed the *Homo negotiatus*, stating that our species is prone to bargain and dispute the value of things until some agreement is reached. The point is that this profile does not erase the fact that negotiation is not necessarily fair since it depends on the parties' starting wealth—or relative power—condition. The basic condition of relative power within social species underlies every animal species' "bargain" or imposing (dominant) behavior. This continuity of power domain—albeit with species-specific profiles and complexities—suggests a *universal evolutionary drive* across species. On this issue, it seems appropriate to mention Price and Feinman (2010) reasonable concern:

> Many aspects of our human condition evolved in the deep past and cannot be fully understood without the long vantage point of history and prehistory. This is certainly true for the fundamental principles of human organization –the structure and function of the operation of society–which have been present for thousands of years. The study of inequality is essentially a concern with the evolution of human society.

Additionally, it seems opportune to include an evolutive puzzle summarized by Fashing (2001):

> … (*Boehm*) notes that both African great apes and humans living in complex societies form dominance hierarchies but that mobile hunter-gatherers adopt an egalitarian lifestyle in which there is little or no difference in political power between individuals. This phenomenon is perplexing because it suggests that political hierarchy followed a U-shaped trajectory during human evolution (Knauft, 1991): from a hierarchical common ancestor with the

[4] Richard C. Wilson, https://www.investopedia.com/ask/answers/08/homo-economicus.asp

African great apes through an egalitarian hunter-gatherer stage to the hierarchies represented by chiefdoms and other more complex civilizations.

This concerning evolutive social dynamic poses one additional question in terms of desirable societal population size. Should humans consider returning to small conglomerates in which egalitarianism could be more feasible? Would these divert the dominance drives to in-between group relationships?

Hierarchical Dominance/Prevalence and Violence

Cummins (2000) refers to dominance as a social reasoning strategy that provided fundamental roots to our cognitive architecture and shaped the human mind and society. As discussed above, this drive for dominance permeated orders and species on evolutionary grounds; it is therefore profoundly entrenched in our evolutive ancestry. As stated in Cummins (2000):

> … social dominance is the earliest stable dimension of peer group social organization (Strayer and Trudel 1984).

According to Boyce (2004), inequalities in human social groups contribute to the initiation and perpetuation of violence. Such inequalities are not only expressed in violations of equal rights in adults but take a significant origin in poor and marginal social settings during the early years of life, affecting emotional and cognitive development (Colombo 2007, 2010, 2019; Lipina and Colombo 2009).

Cummings (2000) rightfully stresses the impact of direct and indirect forms of dominance, through educational, political, and financial means on shaping the human mind and social evolution:

> …our basic cognitive functions were forged in the crucible of dominance hierarchies and were crucial to survival in these environments. Our evolutionary environment is reflected in our cognitive architecture, and that architecture in turn shapes the nature of our social institutions.

As analyzed in previous chapters, dominance/prevalence represents one ancestral drive in animal evolution and manifests among humans in various expressions, whether physical, institutional, or cultural (material, virtual, or religious). Even in egalitarian societies, dominance is considered the expression of a reverse dominance hierarchy by some authors. This form of social structure would be characteristic of small groups and would not survive large population numbers, as discussed in Ames (2010). As mentioned previously (Colombo 2021a), our species occasionally transformed dominance into forms of cruelty, a sociopathic derivation of prevalence.

In evolutive terms, group consolidation is based on interactive factors involving cooperation and dynamic hierarchies, reward, and punishment behaviors. The latter—on neurobiological domains—would depend on the role played by the prefrontal cortex and the mesocortical limbic reward circuitry in cognitive control, namely, the inhibition of a prepotent response to behave selfishly in norm compliance and enforcement (Hayden and Villeneuve 2010). According to these authors, this behavior is enhanced in *H. sapiens*, in which large-scale cooperation among unrelated individuals represents a defining signature of *H. sapiens* culture.

Not germane to the above-described behavior is what Rochat (2005) labels as the profile of Homo negotiatus:

> Social animals need to share space and resources, whether sexual partners, parents, or food. Humans, however, are unique in the way they share as they evolved to become *Homo negotiatus*; a species that is prone to bargain and to dispute the value of things until some agreement is reached.

Inequalities within and among human social groups contribute to the initiation and perpetuation of dominance and violence. Since social inequalities and interpersonal aggression emerge in early life, as Boyce (2004) stated, it is necessary to study their developmental origins in young children's collective and individual behavior. This author concludes that:

> … the prevention of youth violence -and other forms of early disorder- will require … a stronger accounting of how early social inequalities set trajectories toward healthy or disordered behavior.

When Survival Implies Cooperative Dominance Strategies

Are there any ancient evolutive predecessors of a social power struggle and Machiavellism?[5]

In the following statement, can similarities be found with social conditions in the complex history of our human communities?

> For social insects with larger colonies, queen dominance is often replaced by other forms of control. First, there is nutritional coercion. Poorly fed females become small workers and well-fed ones become large queens. This limits the ability of workers to reproduce, but in most species, it does not eliminate it fully. Given an opportunity, workers often will lay eggs. In a large colony, the queen could not successfully police all such behaviour and often ignores it. Instead, other workers do the policing, destroying the eggs of their co-workers. (Queller 2006)

[5] In the field of personality psychology, Machiavellianism is a personality trait centered on manipulativeness, callousness, and indifference to morality. "There is substantial' confirmation of Machs' cynical worldview, pragmatic ethics, and use of duplicitous tactics." (Jones and Paulhus, 2009)

References

Ames KM. On the evolution of the human capacity for inequality and/or egalitarianism. Chapter 2. "Pathways to Power. New Perspectives on the Emergence of Social Inequality". Eds. Price TD, Feinman GM, Springer C. Science & Business Media, LLC 2010.

Beyin A. The Bab al Mandab vs the Nile-Levant: an appraisal of the two dispersal routes for early modern humans out of Africa. Afr Archaeol Rev. 2006;23:5–30. https://doi.org/10.1007/s10437-006-9005-2.

Beyin A. Upper pleistocene human dispersals out of africa: a review of the current state of the debate. Int J Evol Biol. 2011;Article ID 615094 https://doi.org/10.4061/2011/615094.

Boehm C. Egalitarian behavior and reverse dominance hierarchy. Curr Anthropol. 1993;34:227–40.

Boehm C. Conflict and the evolution of social control. J Conscious Stud. 2000;7:79–101.

Boyce WY. Social stratification, health, and violence in the very young. Ann N Y Acad Sci. 2004;1036:47–68. https://doi.org/10.1196/annals.1330.003.

Boyd R, Richerson PJ. Culture and the evolutionary process. London: Univ. Chicago Press; 1988.

Colombo JA. Pobreza y desarrollo infantil (Poverty and child development). Buenos Aires: Ed. Paidos; 2007.

Colombo JA. Somos la especie equivocada? (Are we the wrong species?). Buenos Aires: Ed. EUDEBA; 2010.

Colombo JA. Our Animal condition and social construction. New York: Nova Science Publishers; 2019. ISBN: 978-1-53615-357-6. eBook ISBN: 978-1-53615-3583

Colombo JA. The Homo within the sapiens. New York: Nova Science Publishers; 2021a.

Colombo JA. Ancestral drives and the complex behavior of our species. Int J Primatol Res. 2021b;19:18–9.

Colombo F, Kokkonen V. Man, Matter Metamorphosis. 10.000 years of Design. Berlin, . ISBN 978 951 616 298 3: Finnish Heritage Agency; 2018.

Cooper GA, West SA. Division of labour and the evolution of extreme specialization. Nat Ecol Evol. 2018;2:1161–7.

Cummins DD. How the social environment shaped the evolution of mind. Synthese. 2000;122:3–28.

Etler D. Homo erectus in East Asia: Human ancestor or evolutionary dead end? Athena Rev. 2005;4:37–50.

Fashing PJ. Egalitarianism and group selection in human evolution. Curr Anthropol. 2001; https://doi.org/10.1086/323823.

Hayden B, Villeneuve S. Chapter 5. Who benefits from complexity? A view from futuna. In: Price TD, Feinman GM, editors. Pathways to power, fundamental issues in archaeology. New York: Springer Science+Business Media LLC; 2010. https://doi.org/10.1007/978-1-4419-63000.1. https://link.springer.com/bookseries/5972.

Henrich J, Boyd R. Division of labor, economic specialization, and the evolution of social stratification. Curr Anthropol. 2008;49:71724.

Lahr MM, Foley R. Multiple dispersals and modern human origins. Evol Anthropol. 1994;3:48–60.

Lipina SJ, Colombo JA. Poverty and brain development during childhood. Washington, DC: American Psychological Association; 2009.

Nowak MA. Five rules for the evolution of cooperation. Science. 2006;314:15601563. https://doi.org/10.1126/science.1133755.

Price TD, Feinman GM. Chapter 1. Social inequality and the evolution of human social organization. In: Price TD, Feinman GM, editors. Pathways to power, fundamental issues in archaeology. New York: Springer Science+Business Media, LLC; 2010. https://doi.org/10.1007/978-14419-6300-0_1. https://link.springer.com/bookseries/5972.

Queller DC. To work or not to work. Nature. 2006;442:42–3.

Ratnieks FLW, Helantera H. The evolution of extreme altruism and inequality in insect societies. Philos Trans R Soc B. 2009;364:3169–79. https://doi.org/10.1098/rstb.2009.0129.

Rochat P. Humans evolved to become Homo negotiatus … the rest followed. Behav Brain Sci. 2005;28:714–5. https://doi.org/10.1017/S0140525X05490123.

Scally A, Dutheil JY, Hillier LW, Jordan GE, Goodhead I, et al. Insights into hominid evolution from the gorilla genome sequence. Nature. 2012;483:169–75.

Schmelz M, Grueneisen S, Kabalak A, Jost J, Tomasello M. Chimpanzees return favors at a personal cost. PNAS. 2017;114:7462–7.

Schuster R. Why not chimpanzees, lions, and hyenas too? Behav Brain Sci. 2005;28:716–7.

Segretin MS, Lipina SJ, Hermida MJ, Sheffield TD, Nelson JM, Espy KA, Colombo JA. Predictors of cognitive enhancement after training in preschoolers from diverse socioeconomic backgrounds. Front Psychol. 2014;5:205. https://doi.org/10.3389/fpsyg.2014.00205.

Stout D, Hecht EE. Evolutionary neuroscience of cumulative culture. PNAS. 2017;114:7861–8. https://doi.org/10.1073/pnas.1620738114.

Stout D. Tool Use by Non-Human Primates. Impact of Tool Use and Technology on the Evolution of the Human Mind. Public Symposium. UC San Diego-Salk. October 12, 2018.

Tomasello M. The normative turn in early moral development. Hum Dev. 2018;61:248–63. https://doi.org/10.1159/000492802.

Tomasello M, Carpenter M, Call J, Behne T, Moll H. Understanding and sharing intentions: the origins of cultural cognition. Behav Brain Sci. 2005;28:675–735.

Warneken F, Hare B, Melis AP, Hanus D, Tomasello M. Spontaneous altruism by chimpanzees and children. PLoS Biol. 2007; https://doi.org/10.1371/journal.pbio.0050184.

Whitehead H, Laland KL, Rendell L, Thorogood R, Whiten A. The reach of gene-culture coevolution in animals. Nat Commun. 2019;10:2405. https://doi.org/10.1038/s41467-019-10293-y.

Chapter 8
Population Growth and the *Collective Brain*

> Comparisons between human children and other animals... continue to overlook effects of behavioral and social environment. Such experiments compare chimpanzees that live in cages – lucky if they have a rubber tire to play with or a rope to swing from – with human children that live in the rich environment of suburban homes.
>
> Modern evolutionary and developmental biology (evo-devo) demands that evolutionary theorists recognize the powerful contribution of developmental conditions. (Gardner 2005)

> …profound human characteristics, including but not limited to sociality, are acquired at an early age, while social interactions provide key wiring instructions that determine brain development. (Atzil et al. 2018)

To place matters into an evolutive perspective:

> Humans share many elements of their anatomy and physiology with both gorillas and chimpanzees, and our similarity to these species was emphasized by Darwin and Huxley in the first evolutionary accounts of human origins[1]. Molecular studies confirmed that we are closer to the African apes than to orangutans, and on average closer to chimpanzees than gorillas[2]. (Scally et al. 2012)

Considering such statement, from what genetic-environmental *alchymia* did the human nature emerged? Whatever it was, we carry the main neurobehavioral ingredients from our animal origins, and they permeate our social constructions and dynamics.

Besides socioecological domains to which each species adapt and their genetic composition, there are basic behavioral drives that are constant across the animal kingdom. They include survival (individual and collective), territoriality, feeding, and reproduction. These drives have been considered from insects to primates through the previous chapters if we schematically draw extremes of the brain and social-behavioral organization within the animal kingdom.

As we considered sociality in various animal species, the dimension of social plasticity appeared as an additional character of their social profile. On these grounds, the history of human civilization describes social structures and cultural profiles in terms of adaptational phenotypes, hierarchies, values, and priorities. This same variety suggests a significant degree of plasticity, subjected to dynamic

J. A. Colombo, *Dominance Behavior*, https://doi.org/10.1007/978-3-030-97401-5_8

interactions among basic tenets of natural history (territorialism, survival, dominance) which impinge upon cultural history, demands, and developmental profiles. The emotional dimension plays a significant role in attraction and repulsion modes and tints our behaviors, goals, and desires.

Across animal orders and species, dominance/prevalence and social hierarchies establish feeding access and reproductive rights. Among human social status, they strongly predict well-being conditions, level of education, morbidity rate, and even survival rates. In turn, this affects social integration, early cognitive development, access to life quality, and stressful conditions in subordinated and marginal groups in stable social hierarchies. In dynamic social hierarchies, increased competition and instability would expand stressful conditions to dominant and lower candidates' positions as well.

On neurobiological grounds, applying functional magnetic resonance procedures and reward interactive games mimicking social settings, Zink et al. (2008) observed brain activity changes depending on perceived social rank and stable or unstable hierarchy settings. Brain areas involved included the dorsolateral prefrontal cortex, amygdala, and medial prefrontal cortex.

Of utmost significance is the impact of developmental conditions through the period before attaining individual adolescence. As put forward by Mueller et al. (2013):

> The human brain is characterized by striking inter-individual variability in neuroanatomy and function (Frost and Goebel, 2012; Rademacher et al. 2001; Sugiura et al. 2007; Van Essen and Dierker, 2007) that is reflected in great individual differences in human cognition and behavior. Such variability is a joint output of genetic and environmental influences that may differentially impact on different brain systems (Glendenning and Masterton, 1998).

The mentioned variability places a formidable pressure on the construction of viable ideological and institutional projects.

Behavioral Consequences of Population Density and Population Dynamics (the *Behavioral Sink* of Calhoun). Potential Implications of Calhoun's Experiments

In addition to species-specific sets of behavioral profiles and social structure and interactions, population growth represents a source for dynamic development and social structure and fate outcome. Depending on the considered species, a few alternatives emerge to decrease social entropy with increasing population densities. Besides egalitarian communities, these alternatives include the emergence of hierarchical structures, the process of fission-generating migratory groups, and a potentially degrading social process, as reflected in the experimental setups by Calhoun (1962a, b, 1973) (Fig. 8.1). In addition to the considerations made in a previous chapter, the consequences of overpopulation in these experiments (under spatially

Fig. 8.1 Bagley, Hill, and Calhoun look over the mouse enclosure. Image #04, no date. (National Library of Medicine, John B. Calhoun papers, 1909–1996. MS C 586. Series VII: Negatives, Photographs, and Slides, 1960–1992, box 142, folder 28) (Source: U.S. National Library of Medicine) (https://medicineonscreen.nlm.nih.gov/portfolio/1770/)

limited conditions, but with free-feeding access) are summarized in the following quotations:

> When a population of laboratory rats is allowed to increase in a confined space, the rats develop acutely abnormal patterns of behavior that can even lead to the extinction of the population. (Calhoun 1962a)

> Once the number of rats in a room increased above a certain level, this frequency of contact while eating increased sufficiently that the rats developed a new definition of the feeding situation to include the presence of another rat. Gradually eating in the other three pens declined until 60–80 per cent of all food consumption was in this one of the four pens. The development of this atypical aggregation under the influence of the several conditions and processes involved forms what I have termed a *behavioral sink*.
>
> On the behavioral side, males developed a pansexuality in which they would mount other rats regardless of their age, sex, or receptivity. … the development of a behavioral sink leads to a state of sustained inordinate aggregation which may be called 'pathological togetherness. (Calhoun 1962b)

> If opportunities for role fulfilment fall far short of the demand by those capable of filling roles, and having expectancies to do so, only violence and disruption of social organization

> can follow. Individuals born under these circumstances will be so out of touch with reality as to be incapable even of alienation. Their most complex behaviours will become fragmented. Acquisition, creation and utilization of ideas appropriate for life in a post-industrial cultural conceptual technological society will have been blocked. (Calhoun 1973)

Besides the impact of a territorially distorted relative population distribution, in 1988, Kelley summarized the world population growth conundrum:

> We are living in an unusual era in demographic history. For thousands of years, world population grew at a snail's pace, so slowly, in fact, that it took over 1 million years for population to reach 1 billion-and that was 200 years ago. But the pace quickened, and in a relatively short span of 120 years the population doubled to 2 billion. The third billion took only 35 years to arrive, and fourth, 15. Present world population exceeds 5 billion and demographer do not forecast a levelling until the end of the next century.

Twenty-three years after Kelley's forecast, we have reached almost 8 billion, according to the United Nations Population Division of the Department of Economic and Social Affairs.[1] Reportedly, the worldwide population will be about 9.7 billion by 2050 and slowly reach 10.8 billion by 2100. According to Roser:[2]

> The UN projects that the global population increases from a population of 7.7 billion in 2019 to 11.2 billion by the end of the century. By that time, the UN projects, fast global population growth will come to an end.

Besides absolute world numbers, the issue seems to underline a complex problem of unequal distribution—which Kelley focalizes in the Third World—and worse, a problem of world inequality regarding access to a humanly quality of life. This growth is unevenly distributed among regions, nations, and socioeconomic standings. The impact of this issue is summarized on fertility and death rates, local economies, personal wealth, education distribution (both blatantly unequal), and available food production and healthcare technologies. Treatment of each of them exceeds the scope of this book. They converge on individual and collective quality of life, social aggression and institutional stability, and emerging social pathologies or deviant behaviors. The experiment outlined above somehow represents a vivid metaphor of social and individual degradation under experimental overpopulation conditions in enclosed settings. In the real human world, such an "enclosure" is represented by local economic conditions and distributed access to education and food resources, poverty, living quarters crowding, and social marginalization. Also, ecological abuse and degradation and lack of recovery processes will impact the mentioned "enclosure" effect. Hence, Calhoun's experiments, albeit in laboratory-controlled conditions, do not seem to represent an impossible metaphor of real life among existing living inequalities and its impact on human behavior basic drives and possible socio-pathological outcomes.

Though analyzing human population growth and potential outcome in the long range is beyond the scope of this book, it merits to mention a dreadful view on

---

[1] https://www.worldometers.info/world-population/#:~:text=7.9%20Billion%20 (2021)

[2] https://ourworldindata.org/future-population-growth

human population dynamics expressed by Gee (2021) (editor of the scientific journal *Nature*) in his article *Humans are doomed to go extinct.* In it he summarizes that habitat degradation, low genetic variation, and declining fertility are setting *Homo sapiens* up for collapse. According to the author, current human population has grown very rapidly from a reduced number of original members which results in reduced genetic variation, a negative prospect for any species survival.

Experiments based on enriched environments represent the other side of the coin. Such experiments, based on the pioneer works of Levine (1959), Bennett et al. (1969), Greenough et al. (1973), Kempermann et al. (1997), and Soffie et al. (1999), and later by Jung and Herms (2014), show a direct impact of physical and social enrichment on basic factors of microstructural neuronal-glial development and connectivity in several brain structures and behavior. It also impinges upon neuroplasticity and cognitive and brain reserves, as Petrosini et al. (2009) described. According to Jung and Herms (2014):

> ... we speculate that the cognitive benefits seen in environmental-enriched animals might be a consequence of both, a higher connectivity of the neuronal network due to more established synapses and an enhanced flexibility due to more transient spines.

According to Fang and Yuste (2017):

> ...variations in neuronal density not only occur across cortical areas in the same species (Collins et al. 2010), but also within the same cortical region among individuals (Herculano-Houzel, 2009). The intraspecific variability in neuronal number probably influences the structural and functional basis of neural circuits, leading to differences in brain function and behavior (Mueller et al. 2013; Song et al. 2013; Vogel and Machizawa, 2004; Ward et al. 1998).

In terms of an enriched environment, the parental role should be stressed, as it was earlier shown by Harlow (1959) and more recently by Uller (2012):

> ... mothers and fathers have a significant impact on their offspring well beyond birth in many species through resource provisioning and behavioural interactions, which play important roles in the ontogeny of species-typical phenotypes and individual phenotypic variation (Gottlieb 1997; Avital and Jablonka 2000; Maestripieri 2009; Michel 2011).

Population Growth, the *Collective Brain*, and the Brain Size Factor

The above considerations would lack a more comprehensive view of the concept if the collective brain is avoided. The concept of cultural interactions is undoubtedly affected by the emergence of language, technological developments in communication, and collective interactions. The following concept by Muthukrishna and Henrich (2016)—for whom societies and social networks act as collective brains—perhaps summarizes the role of such interactive process in social and technological evolution:

> …innovations, large or small, do not require heroic geniuses any more than your thoughts hinge on a particular neuron. Rather, just as thoughts are an emergent property of neurons firing in our neural networks, innovations arise as an emergent consequence of our species' psychology applied within our societies and social networks. Our societies and social networks act as collective brains.

The creative domain in the human species was considered in a previous publication (Colombo 2019), as succinctly expressed in the following paragraph:

> … let us aim at *creativity*, a mental process whose expression is maximized in the *anatomically modern H. sapiens*. That is, the capacity for '*divergent thinking*' and '*conceptual expansion*', and their productive expression, either material or virtual…

Regarding the collection and connectivity of processing units in mammals, brain size—and its correlation with body size—has been associated with cognitive demands (Dunbar 2009) and the acquisition of novel, plastic, behavioral strategies in mammalian vertebrates (Shultz and Dunbar 2007). According to the cultural brain hypothesis, brain size and mental processing underwent primary selection pressure for brain size across many taxa, linked to adaptive knowledge and problem-solving (Muthukrishna and Henrich 2016). González-Forero and Gardner (2018), based on a metabolic model to approach possible drives for human brain size evolution, argue supporting social challenges as a component of ecologically driven brain expansion in *Homo*. According to these authors, competition between individuals has been unimportant for driving human brain size evolution, adding that:

> … because our analysis suggests that brain expansion in Homo has not been driven by peer cooperation or competition, our results indicate that social complexity may have had a more limited role in human brain size expansion than is commonly thought.

Larger brains to body size? In insects, advanced behavioral repertoires might require large and sophisticated mushroom bodies. On the contrary, as Mares et al. (2005) discussed, based on research performed in bumblebees and honeybees, increased division of labor may require less, rather than more, neuronal capacity. Individuals of highly social insect species with an advanced division of labor and task specialization might not need to perform the entire behavioral repertoire of the species. According to these authors, depending on the insect species, labor specialization would result in fewer brain demands, evolving as smaller brains in which adult neurogenesis and dendritic outgrowth are deciding factors.

Adaptive use of social information would not require large brains but is achieved through relatively simple social learning strategies, which could be inherited genetically or arise through learning. Hence, according to Gruter and Leadbeater (2014), information-use strategies are more likely to reflect ecological selection pressures than brain size or phylogenetic relationships and suggest that insects and vertebrates use similar information-use strategies. However, comparative studies tend to conform to the idea that larger brains to body size allow animals to generate more elaborate behavioral repertoires or more complex social behaviors. This evolutive issue was analyzed by several authors, and its treatment exceeds the objectives of the present essay.

References

Atzil S, Gao W, Fradkin I, Barrett LF. Growing a social brain. Nat Hum Behav. 2018;2:624–36.

Bennett EL, Rosenzweig MR, Diamond MC. Rat brain: effects of environmental enrichment on wet and dry weights. Science. 1969;163:825–6.

Calhoun JB. Population density and social pathology. Scientific American, Inc. 140.142.214.159. 1962a.

Calhoun JB. Chapter 22: A behavioral sink. In: Bliss EL, editor. Roots of behavior. New York: Harper; 1962b.

Calhoun JB. Death squared. The explosive growth and demise of a mouse population. Proc Roy Soc Med. 1973;66(1 Pt 2):80–8.

Colombo JA.Our Animal condition and social construction. Nova Sci. Publ., USA. 2019. ISBN: 978-1-53615-357-6. eBook ISBN: 978-1-53615-3583.

Dunbar RIM. The social brain hypothesis and its implications for social evolution. Ann Hum Biol. 2009;36:562–72.

Fang W-Q, Yuste R. Overproduction of neurons is correlated with enhanced cortical ensembles and increased perceptual discrimination. Cell Rep. 2017;21:381–92. https://doi.org/10.1016/j.celrep.2017.09.040.

Gardner RA. Animal cognition meets evo-devo. Behav Brain Sci. 2005;28:699–700. https://doi.org/10.1017/S0140525X05490123.

Gee H. Humans are doomed to go extinct. Scientific American nov. 30, 2021. https://www.scientificamerican.com/article/humans-are-doomed-to-go-extinct/

González-Forero M, Gardner A. Inference of ecological and social drivers of human brain- size evolution. Nature. 2018;557:554–7. https://doi.org/10.1038/s41586-018-0127-x.

Greenough WT, Volkmar FR, Juraska JM. Effects of rearing complexity on dendritic branching in frontolateral and temporal cortex of the rat. Exp Neurol. 1973;41:371–8.

Gruter C, Leadbeater E. Insights from insects about adaptive social information use. Trends Ecol Evol. 2014;29:177–84. https://doi.org/10.1016/j.tree.2014.01.004.

Harlow HF. Psychobiology, *Scientific American* 200. San Francisco: W.H. Freeman and Co.; 1959. p. 68–74.

Jung CKE, Herms J. Structural dynamics of dendritic spines are influenced by an environmental enrichment: an in vivo imaging study. Cerebral Cortex. 2014;24:377–84. https://doi.org/10.1093/cercor/bhs317.

Kelley AC. Economic consequences of population change in the Third World. J Economic. 1988;26:1685–728.

Kempermann G, Kuhn HG, Gage FH. More hippocampal neurons in adult mice living in an enriched environment. Nature. 1997;386:493–5.

Levine S. Stimulation in infancy. In: Psychobiology, *Scientific American*. San Francisco: W.H. Freeman and Co.; 1959. p. 93–8.

Mares S, Ash L, Gronenberg W. Brain allometry in bumblebee and honeybee workers. Brain Behav. Evol. 2005;66:50–61. https://doi.org/10.1159/000085047.

Mueller S, Wang D, Fo MD, Yeo BTT, Sepulcreb J, Sabunc MR, Shafee R, Lu J, Liu H. Individual variability in functional connectivity architecture of the human brain. Neuron. 2013;77:586–95. https://doi.org/10.1016/j.neuron.2012.12.028.

Muthukrishna M, Henrich J. Innovation in the collective brain. Phil Trans R Soc B. 2016;371:20150192. https://doi.org/10.1098/rstb.2015.0192.

Petrosini L, De Bartolo P, Foti F, Gelfo F, Cutuli D, Leggio MG, Mandolesi L. On whether the environmental enrichment may provide cognitive and brain reserves. Brain Res Rev. 2009;61:221–38.

Scally A, Dutheil YA, Hillier LW, Jordan GE, Goodhead I. (et al.) Insights into hominid evolution from the gorilla genome sequence. *Nature*. 2012;483:169-175.

Shultz S, Dunbar RIM. The evolution of the social brain: anthropoid primates contrast with other vertebrates. Proc R Soc B. 2007;274:2429–36. https://doi.org/10.1098/rspb.2007.0693.

Soffie M, Hahn K, Terao E, Eclancher F. Behavioural and glial changes in old rats following environmental enrichment. Behav Brain Res. 1999;101:37–49.

Uller T. Parental effects in development and evolution. In: Royle NJ, Smiseth PT, Kölliker M, editors. The evolution of parental care. 1st ed. Oxford University Press; 2012.

Zink CF, Tong Y, Chen Q, Bassett DS, Stein JL, Meyer-Lindenberg A. Know your place: neural processing of social hierarchy in humans. Neuron. 2008;58:273–83.

Chapter 9
Tooling and Technology

> Far from a breakthrough, stone tool-making appears to have been a fragile behavior of marginal value. This is expected for a technology located near the limits of contemporary hominin capacities, with high learning costs, limited flexibility, and high rates of failure offsetting its benefits. Natural selection acting on hominin brains and bodies during this lengthy period of experimentation may eventually have eased these costs, leading to a dramatic proliferation of 'Oldowan' tool-sites after about 2.0 million years ago (mya). This is closely coincident with the first appearance of larger-brained and –bodied *Homo erectus* by ~1.9 mya and was rapidly followed by the invention of 'ore sophisticated Acheulean 'handaxe" technology by 1.76 mya, marking an important inflection point in the biocultural feedback processes that eventually produced the modern human technological niche. (Dietrich Stout 2018)

> Tools are ubiquitous in human cultures and have been central to human lifestyles since the origin of the genus Homo (Gibson and Ingold, 1994). (Fragaszy and Mangalam 2018)

Tool Use, Material Culture, Tooling, and the Emergence of Technology

The following statements illustrate the conceptually dynamic issue that represents the evolutive analysis of tool use and manufacture across species and its conceptual definition:

> The definition of 'tool use' is problematic, often arbitrary or subjective, sometimes anthropocentric, and open to interpretation. (Bentley-Condit and Smith 2010)

> A recent, comprehensive compendium on non-human animal tool-use behaviour [1] lists four phyla (Echinodermata, Arthropoda, Mollusca and Chordata) and nine classes (sea urchins, insects, spiders, crabs, snails, octopi, fish, birds and mammals) as containing tool-using species…. This broad phylogenetic spread of multiple origins, however, goes hand in hand with overall rarity: tool use has been documented in less than 1% of the animal genera currently identified, and an even smaller percentage of species. The evolutionary events that gave rise to this eclectic distribution must find their ultimate explanation in the benefits that tool use offers to individuals in these species. (Biro et al. 2013)

J. A. Colombo, *Dominance Behavior*, https://doi.org/10.1007/978-3-030-97401-5_9

While the difference between tool use and tool manufacture defines their cognitive consequences, differences between "flexible tool users" and "stereotyped tool users" are based on whether mechanical actions are or not intended to have a goal-directed interaction with another object. According to Baber (2003) (cf. Biro et al. 2013), this has an evolutive value. F*lexible tool use* is not phylogenetically widespread and seems to require a certain level of cognitive processing (also discussed in Cabrera-Álvarez and Clayton 2020). As mentioned in Hunt and Uomini (2016), flexible tool use is primarily found in birds and primates. Insects and fish mostly show stereotypical tool use. Whether tool behavior depends exclusively on a species-specific, stereotyped, hereditary code or involves a previous interactive learning and active adaptation process. Cumulative modification of tool designs over time is crucial for the development of technological evolution. Under the dynamics of behavioral evolution, these differences could be envisioned as adaptive stages involving specific interactions based on species cognitive-ecologic contexts.

The emergence of cumulative modified tools represents another evolutive stage toward the dawn of technology, probably based on the ability to perceive causal relationships that could be crucial to the foundation of modern technology (Wolpert 2003). This implies the capacity for an insight into causality interactions among objects and the construction of mental projects based on material representations. On comparative grounds, as stated by Maravita and Iriki (2004) in a relatively soft argumentation:

> ...we could imagine that some precursor (or basic building block) of human tool-use ability was already furnished in the brain of our common ancestor with monkeys and was pushed to full expression by some sort of 'evolutionary pressure'.

Hunt and Uomini (2016) state:

> ... we compared the tool-associated adaptive patterns in Homo erectus and the New Caledonian crow. Both species provide the most convincing early Homo and nonhuman evidence, respectively, for the making of cumulatively modified tools.

According to the above authors, causal relationships between multiple tools under direct or virtual vision involve prefrontal-intraparietal interactions. This could be the basis for developing technological strategies proper for humans (see also Wolpert, 2003; Seed and Byrne 2010). From a neuroscientific approach, it is interesting to mention that Hunt and Uomini (2016) proposed that tools can become incorporated into a plastic neural representation of our body, by enabling us—through internalization—to extend our reach in space.

As stated by Hunt and Gray (2003):

> ...New Caledonian crows therefore provide a unique opportunity to identify the cognitive and neural features that are required for sophisticated tool manufacture and cumulative technological evolution. The cumulative nature of natural selection is an essential requirement for adaptive biological evolution (Dawkins 1988). A similar cumulative process in tool design has allowed humans to develop progressively more sophisticated tool lineages; the evolution of the axe is a good example (Oakley 1961). This cumulative change is a 'ratchet-like' process where design changes are retained at the population level until new improved designs arise (Tomasello et al. 1993; Tomasello 1999).

The evolution of material culture among *Homo* would be represented by the last phases of human prehistory, after the emergence of *Homo sapiens*. In this context, it is interesting to include the concept of Colombo and Kokkonen (2018) framed in the following statement:

> The equation Man Matter Metamorphosis is a primal and constant aspect of human-ecosystem interaction… The universe of artifacts that results from this equation is a constitutive element of societies.

This integrated concept relates to the ecological framework of "tooling," in keeping with an ethological approach to behavior. This issue was extensively developed by Fragaszy and Mangalam (2018), in terms of differentiating "tool use" with the motion sequence of tooling. The latter involves concepts based on action theory, space perception and cognition, and biomechanics. As stated by Fragaszy and Mangalam (2018):

> An individual that is tooling is referred to hereafter as a 'tooler.' Thus, we propose a theory of tooling. Because our theory of tooling is framed in biomechanical and spatial concepts of action, currently popular descriptive definitions of 'tool' and 'tool use' are not relevant to it.
>
> *Tooling* is deliberately producing a mechanical effect upon a target object/surface by first grasping an object, thus transforming the body into the body-plus-object system, and then using the body-plus-object system to manage (at least one) spatial relation(s) between a grasped object and a target object/surface, creating a mechanical interface between the two.

Thus, further advances in the analysis of tool use and manufacture definitions are left to other specific discussion sources. Here it will be limited to its application among primates concerning *Homo* and its development toward the concept of technology and its impact on social structure and dynamics. In this context, it seems that the following statement by Haslam (2014) defines the probable timing on a primate evolutionary scale:

> Here, I use available data on wild bonobo tool use and emerging molecular estimates of demography during Pan evolution to hypothesize the plausible tool use behavior of the bonobo-chimpanzee Last Common Ancestor (LCA) at the start of the Pleistocene, over 2 million years ago. This method indicates that the common ancestor of living Pan apes likely used plant tools for probing, sponging, and display, but it did not use stone tools. Instead, stone tool use appears to have been independently invented by Western African chimpanzees (P. t. verus) during the Middle Pleistocene in the region of modern Liberia-Ivory Coast-Guinea, possibly as recently as 200,000-150,000 years ago. If this is the case, then… this trait probably first emerged among hominins in Pliocene East Africa.

Stone tool development by *Homo* was summarized in Colombo (2020 a), considering further technological development implications in social interactions and development. Material manipulation by personal crafting or applying combined technologies that displace direct contact with the material product hints for a possible evolution of individual material manipulation and object creation toward machine-mediated production according to designer strategies. Perhaps the concept of crafting approaches the modern design concept, with the difference of a direct material contact in the first case, i.e., the cultural evolution of technology would underscore the collective development of successive stages in craftmanship.

As stated in Boyd et al. (2013):

> People in even the simplest human societies depend on tools that are beyond the inventive capacity of individuals.

From an evolutive neuroscientific perspective, it seems interesting to include here comparisons made on tool use between corvids and apes to analyze the convergent evolution in those animal orders. As stated by Emery and Clayton (2004):

> There are many aspects of corvid and ape cognition that appear to use the same cognitive tool kit: causal reasoning, flexibility, imagination, and prospection. We suggest that nonverbal complex cognition may be constructed through a combination of these tools. Although corvids and apes may share these cognitive tools, this convergent evolution of cognition has not been built on a convergent evolution of brains. Although the ape neocortex and corvid nidopallium are both significantly enlarged, their structures are very different, with the ape neocortex having a laminar arrangement and the avian pallium having a nuclear arrangement (2). It is unclear what implications these structural differences have. However, **cognition in corvids and apes must have evolved through a process of divergent brain evolution with convergent mental evolution.** This conclusion has important implications for understanding the evolution of intelligence, given that it can evolve in the absence of a prefrontal cortex. (Bold characters from author JAC)

Finally, as stated by Wolpert (2003):

> It is proposed that the evolution of causal thinking was essential for the development of tool use, as it is not possible to make a complex tool without understanding cause and effect.

On cognitive grounds, regarding the analysis of the evolutive, comparative, sense of quantity, the following concepts (Barras 2021) succinctly describe the issue:

> …researchers once thought that humans were the only species with a sense of quantity, studies since the mid-twentieth century have revealed that many animals share the ability. For instance, fish, bees and newborn chicks[3] can instantly recognize quantities up to four, a skill known as subitizing. Some animals are also capable of 'large-quantity discrimination': they can appreciate the difference between two large quantities if they are distinct enough. Creatures with this skill could, for example, distinguish 10 objects from 20 objects, but not 20 from 21. Six-month-old human infants also show a similar appreciation of quantity, even before they have had significant exposure to human culture or language. What all of this suggests, says Andreas Nieder, a neuroscientist at the University of Tübingen, Germany, is that humans have an innate appreciation of numbers. That arose through evolutionary processes such as natural selection, he says, because it would have carried adaptive benefits. (*Which implies cultural evolution*, JAC). Rafael Núñez, a cognitive scientist at the University of California, San Diego, suggests that a distinction should therefore be made between what he has dubbed the innate 'quantical' cognition seen in animals and the learnt 'numerical' cognition seen in humans[2].

References

Barras C. How did ancient humans learn to count? Nature. 2021;594:22–5.
Bentley-Condit VK, Smith EO. Animal tool use: current definitions and an updated comprehensive catalog. Behaviour. 2010;147:185–221.
Biro D, Haslam M, Rutz C. Tool use as adaptation. Philos Trans R Soc B. 2013;368:20120408. https://doi.org/10.1098/rstb.2012.0408.
Boyd R, Richerson P, Henrich J. The cultural evolution of technology: facts and theories. Chapter 7. In: Richerson PJ, Christiansen MH, editors, Lupp J, series editor. Cultural evolution: society, technology, language, and religion. Strüngmann Forum Reports, vol. 12. Cambridge, MA: MIT Press; 2013. p. 119–42. ISBN 978-0-262-01975-0.
Cabrera-Álvarez MJ, Clayton NS. Neural processes underlying tool use in humans, macaques, and corvids. Front Psychol. 2020;11:560669. https://doi.org/10.3389/fpsyg.2020.560669.
Colombo JA. Creativity, a profile for or species. Newcastle, UK: Cambridge Scholars Publishers; 2020.
Colombo F, Kokkonen V. Man, matter metamorphosis. 10.000 years of design. Berlin., ISBN 978 951 616 298 3: Finnish Heritage Agency; 2018.
Emery NJ, Clayton NS. The mentality of crows: convergent evolution of intelligence in corvids and apes. Science. 2004;306:1903–7.
Fragaszy DM, Mangalam M. Tooling. In: Advances in the study of behavior. San Diego: Elsevier Inc.; 2018. ISSN 0065-3454.
Haslam M. On the tool use behavior of the bonobo-chimpanzee last common ancestor, and the origins of hominine stone tool use. Am J Primatol. 2014;76:910–8. https://doi.org/10.1002/ajp.22284.
Hunt G, Gray RD. Diversification and cumulative evolution in New Caledonian crow tool manufacture. Proc R Soc Lond B. 2003;270(867–874):867. https://doi.org/10.1098/rspb.2002.2302.
Hunt GR, Uomini N. A complex adaptive system may be essential for cumulative modifications in tool design. Jpn J Anim Psychol. 2016;66:141–59.
Maravita A, Iriki A. Tools for the body (schema). Trends Cogn Sci. 2004;8:79–86.
Seed A, Byrne R. Animal tool-use. Curr Biol. 2010;20(23):R1032–9.
Stout D. Tool use by non-human primates. Impact of tool use and technology on the evolution of the human mind. Public symposium. UC San Diego-Salk. October 12, 2018.
Wolpert L. Causal belief and the origins of technology. Philos Trans R Soc A. 2003;36:1809. https://doi.org/10.1098/rsta.2003.1231.

Chapter 10
The *Homo sapiens*, Evolution of the Warmonger Human

Clutton-Brock (2009) refers to a statement by C. Darwin, construed in another conceptual and cultural context, related to morality and conscience in animal life:

> In *The descent of man* (1871), Darwin turned to the evolution of human societies. In Chapter VI, he stresses the contrast between humans and other animals 'I fully subscribe to the judgement of those writers who maintain that of all the differences between man and the lesser animals, the moral sense or conscience is by far the most important' (*The descent of man*, p. 97).

Some of those concepts have been tapped in the previous and the following paragraphs, pursuing a comparative, evolutive view.

Intelligence is an evolutionary adaptation affected by social complexity, social size grouping, and neural circuits complexity. In evolutive domains, Machiavellian intelligence evolves mainly within those parameters, also expressed in nonhuman primates (Whitten and Byrne 1997). However, behavioral adaptations in non-primate species manifest different strategies, suggesting that neurobehavioral plasticity conceals forms of evolutionary Machiavellism as a means for adaptive and dominance/prevalence.

Examples of brain-behavioral changes abound among social animals (see previous chapters). Besides the pursuit of power or social benefits in nonhuman primates and human species, perhaps one bizarre case is represented among insects by a socially flexible ant (Harpegnathos saltator) (Penick et al. 2021):

> Unlike most ant species, workers of H. saltator are capable of sexual reproduction, and they compete in a dominance tournament to establish a group of reproductive workers, termed 'gamergates'.

Various species' behavior provides examples of plasticity of brain mechanisms involved in power access, among others, competence in sex reversal in subordinate females of a coral reef fish following the death of the dominant male (Robertson 1972) and brain changes in songbirds for reproductive and territorial purposes, most notable in animals that occupy seasonal environments (Tramontin and Brenowitz 2000) (based on the work by Nottebohm et al. 1986). Hence, neural and behavioral plasticity, oriented toward social dominance/prevalence or survival, or reproductive goals exist in the natural kingdom at various organismal levels (e.g., neural reorganization). Some species can express phenotypic reversibility, though it occurs in counted cases because it has limitations and costs according to Penick et al. (2021).

J. A. Colombo, *Dominance Behavior*, https://doi.org/10.1007/978-3-030-97401-5_10

Returning to human and nonhuman primate species, Machiavellism or behavioral camouflaging objectives prioritize the goal over the means, implementing deceit and social manipulation (Whitten and Byrne 1997). It represents one ancient cultural strategy to survive, avoid punishment, acquire power, or gain social advantage. Within this context, disclosing the fundamental goals and intentions of political leaders and corporative interests represent one actual conundrum for worldwide citizens and a source of voluntary or involuntary submission, resistance, and uncertainty.

Our Species Must Deal with Evolutive Components, the Sources of *Social* and Developmental Inequalities, and the Hidden and Open Violence It Engenders

> …we now predict that a greater fraction (around 5.1%) of the human genome is genetically closer to chimpanzee or bonobo compared to previous studies (3.3%)[1].
>
> Nicholas Humphrey's (1976) social intelligence hypothesis proposed that the major engine of primate cognitive evolution was social competition. This competitive aspect was emphasized even further in similar later proposals espousing 'primate politics' (de Waal 1982) and Machiavellian intelligence (Byrne and Whiten 1988) … at about the same time as Humphrey's original proposal, Lev Vygotsky's (1978) general theory of culture first appeared in English translation…Vygotsky also emphasized the social dimension of intelligence, but he focused on cooperative things… (Mao et al. 2021)

Results of the interaction of these opposing drives would depend on their relative prevalence and whether are expressed as inter- or intragroup social domains. Both generate behavioral derivates, such as conquest and dominance/prevalence or constructive interactions, thus providing grounds for mixed genetic breeding. This took place among genetically, territorially, and culturally related and unrelated phenotypes. It would thus appear that both competition and cooperation—strong reciprocity—have fueled basic processes that have taken place as fission-fusion social dynamics, promoting group dispersion and cohesion.

In modern times, humans' unequal access to cognitive and technological development has provided grounds for furthering the social gap within and among communities and countries. Prevalence of competition—based on the concept of physical or comparative standing in terms of self-survival condition—has often resulted, besides dominance or conquest, in what can be considered individual and collective sociopathic behaviors. These, in turn, have promoted arm race proliferation, dominance/prevalence-prone development in terms of technologies, and mass promotion strategies favoring the spread of the cultural gap between collective awareness and information access. The result has been the promotion of instruments for dominance and conversion of inequality into dominance/prevalence. Hence, free-running competitiveness has proven to be a source of an actual crackling of the social network. Also, it has been an inducer of inegalitarian access to knowledge and quality of life for vast human groups and a hazard for the environment.

As mentioned in previous chapters, social inequality merged under diverse socio-ecological conditions and would represent a universal component in the animal kingdom. Must our species passively proceed with such a backpack or, instead, attempt a cultural transformation?

Tomasello et al. (2005) proposed that the crucial difference between humans' and other species' cognition resides in the ability to participate in collaborative activities with shared goals and intentions: shared intentionality. Besides the psychological and behavioral differences in generating a virtual complex of values and symbols supporting cultural development, the uniqueness of shared intentionality attributed to humans has not gone unchallenged, as previous chapters may suggest. More blatantly stated by Boesch (2005):

> Ignoring most published evidence on wild chimpanzees, Tomasello et al.'s claim that shared goals and intentions are uniquely human amounts to a faith statement. A brief survey of chimpanzee hunting tactics shows that group hunts are compatible with a shared goals and intentions hypothesis. The disdain of observational data in experimental psychology leads some to ignore the reality of animal cognitive achievements.

Also, by Gardner (2005):

> Sound comparative psychology and modern evolutionary and developmental biology (often called evo-devo) emphasize powerful effects of developmental conditions on the expression of genetic endowment. Both demand that evolutionary theorists recognize these effects. Instead, Tomasello et al. compares studies of normal human children with studies of chimpanzees reared and maintained in cognitively deprived conditions, while ignoring studies of chimpanzees in cognitively appropriate environments.

As well as by Schuster (2005):

> Examples are cited of group hunting in chimpanzees, lions, and hyenas consistent with evidence for intentionality, organization, and coordination. These challenge the claim for shared intentionality as uniquely human. Even when rarely performed in this way, the significance of such behaviors should not be minimized, especially if this level of 'intelligent' action emerges spontaneously in the wild.

Besides the fact that there is no consensus concerning which brain states or processes should be considered as reliable signatures of conscious processing, even in terms of the actual existence of consciousness in other species, there are unresolved conceptual disputes. On these grounds, Boly et al. (2013) place an additional viewpoint on the analysis of human differences within the animal kingdom:

> Recent years have witnessed increasing interest in the phylogenetic origins of consciousness (Edelman and Seth, 2009; Feinberg and Mallatt, 2013). Accumulating evidence indicates the presence of complex cognitive processes in animals, such as birds and certain invertebrates that are phylogenetically distant from humans. Examples include the presence of working memory, social learning, planning, and problem solving in birds (Pepperberg, 2006; Salwiczek et al. 2010), as well as the presence of rich behaviors, sophisticated learning and memory faculties in cephalopod mollusks such as the octopus (Edelman and Seth, 2009).

Departure from the animal kingdom in terms of the capacity to generate a complex system of values and symbolisms that laid down our cultural evolution has brought up, once more, a debate of the roles played by genetic and environmental domains. As Richerson and Boyd (2005) have clearly stated:

> Not by genes alone.

This statement further adds to the provocative inquiry on how culture transformed human evolution. According to these authors, three highly interactive factors could act as proximate causes of this variation. These are inheriting different genes, raising in different environments, and acquiring different beliefs, values, and skills through teaching and observational learning. Among humans, the latter is enhanced by collective access to media means that tend to mold public opinions and expectations regarding various issues of social life and supposed material needs.

A further compounding issue involves the size of the social group. According to primate comparative studies:

> Comparative analyses indicating that cooperative territoriality commonly occurs in primate species who live in small groups (5) and field studies demonstrating that individual participation in intergroup encounters decreases with group size (11, 12, 23) have been interpreted as evidence of the difficulty of achieving collective action in large groups. (Langergraber et al. 2017)

Hence, the development of individual opinions on everyday life issues is trapped under several profilers. They include social standing, subconscious comparative self-evaluation, induction of supposed material needs, and social size. Additionally and perhaps the most undesirable factor in human societies is represented by manipulating consciousness or awareness on social and developmental cognitive issues by public media implementing several means, whether appealing to rational or subconscious processing. These affect personal development and state of social alert and their cultural development as active citizens. In this context arises the concept of memes as cultural transmission units.

A field of study called memetics arose in the 1990s to explore the concepts and transmission of memes in terms of an evolutionary model based on transmission units.

> A meme is an idea, behavior, or style that spreads by means of imitation from person to person within a culture and often carries symbolic meaning representing a particular phenomenon or theme. A meme acts as a unit for carrying cultural ideas, symbols, or practices, that can be transmitted from one mind to another through writing, speech, gestures, rituals, or other imitable phenomena with a mimicked theme. Supporters of the concept regard memes as cultural analogues to genes in that they self-replicate, mutate, and respond to selective pressures.[1]

Dawkins defined the meme as a unit of cultural transmission or imitation and replication, though later definitions would vary. The lack of a consistent, rigorous, and precise understanding of what typically makes up one unit of cultural transmission remains a problem in debates about memetics. The word meme itself is a neologism

---

[1] https://en.wikipedia.org/wiki/Meme

coined by Richard Dawkins (1995), originating from his 1976 book *The Selfish Gene*. In the broadest sense, the objects of copying are memes, whereas objects of translation and interpretation are signs.

Impact of Technological Evolution on Our Degrees of Freedom; Neural and Behavioral Domains

Modern human behavior is characterized by increased complexity in symbolic language and technological development. According to Powell et al. (2009):

> ... demography is a major determinant in the maintenance of cultural complexity and variation in regional subpopulation density and/or migratory activity results in spatial structuring of cultural skill accumulation.

Adaptation to the rate of technological advances in communication and its impact on the interaction of the individuals with their environment and sense of future has bogged citizens' minds with a sense of being passively carried to accept or adapt to third parties' design of supposed needs and resources. This blurs the perception of the future, notwithstanding losing the path of its future as a collective social group. In other words, citizens' awareness of detachment from a decision process that affects future social designs and way of life introduced an additional component that affects comprehension and decisions regarding their future and that of the world as we know it.

Besides the collective human sense of power to undergo daring enterprises, a series of technological developments (see further below) stage conditions for the citizens' future without allowing them to recognize developing issues or affect the course of action. Furthermore, they promote loosening citizens' involvement in how entrepreneurs decide measures of collective impact, that is, the direction of financial power, quite often neglecting urgent terrestrial needs in terms of collective development and attaining humane levels of quality of life and equal educational opportunities.

Previous chapters mentioned an overview of adaptations and social behaviors among species as evolutive distant as insects and primates. Among other significative differences is their relative dependence on the ecological environment.

Humans have developed a collective sense that overvalues our goals and desires beyond any reasonable interaction with social inequalities and our ecological environment. The consequences have reached attention in public media. Thus, a hopefully progressive state of collective conscience, wishfully, will manage to curve the consequences of some actions that rather belong to irrational human cultural phenotypes. Among them, profit exploitation of natural reserves—earthbound or not—has relegated priorities for our home planet and fellow citizens.

Besides political power exercises linked to high-technology developments, power standings in social contexts profoundly impact human development and behavior. On a neurobiological basis, Zink et al. (2008) revealed neural systems involved in consequences of the hierarchical status on individual behavior and emotions:

> In both stable and unstable social hierarchies, viewing a superior individual differentially engaged perceptual-attentional, saliency, and cognitive systems, notably dorsolateral prefrontal cortex. In the unstable hierarchy setting, additional regions were recruited related to emotional processing (amygdala), social cognition (medial prefrontal cortex), and behavioral readiness.

It is interesting that hierarchical settings impinge upon individual development since early postnatal days. The transference and transformation of such early potential imprinting on later social hierarchical settings remain to be further analyzed. Furthermore, the concept of interacting hierarchies and personal development has a significant relationship with social unrest and political (ideological) expressions. In this context, financial interests, individual freedom, collective ideals, power strongholds, and social inequities conform to the interactive bases of historical changes in institutional dynamics.

Using functional magnetic resonance imaging (fMRI) in chimeric interactive simulated social contexts, Zink et al. (2008) analyzed dissociable neural responses to perceived social rank. In summary, their results suggest that perceived hierarchical status can be either fixed or changeable, and this aspect of social stratification has pronounced implications for individuals. In this regard, as advanced by Sapolsky (2004, 2005) in primates, the more subordinate position in stable social hierarchies is associated with more significant stress. In contrast, the dominant position experiences the most stressors in dynamic hierarchies due to increased competition and instability.

> One of the greatest challenges in public health is to understand the socioeconomic gradient. This refers to the fact that in numerous Westernized societies, stepwise descent in socioeconomic status (SES) predicts increased risks of cardiovascular, respiratory, rheumatoid, and psychiatric diseases; low birth weight; infant mortality; and mortality from all causes (1–4). (Sapolsky 2005)

Thus, under various circumstances, social dominance/prevalence can be associated with the most stress-related pathology, whereas in other situations, this is a trait of subordinate individuals. As stated by Sapolsky (2005):

> … it is a testimony to the power of humans, after inventing material technology and the unequal distribution of its spoils, to corrosively subordinate its have-nots.

As Sapolsky (2004) also stated:

> The developments of class, stratification, and poverty are fairly recent in hominid history. What these findings suggest is that nothing in the world of nonhuman sociality involves such an utterly, psychologically permeating sense of subordination as does the human invention of poverty.

Besides the unrefutable statement about poverty—as reviewed in other circumstances—this statement's initial argument could be challenged since, as mentioned in previous chapters, inequality is expressed among members across the animal kingdom in terms of differential rights to food and reproductive access. However, the point to be stressed here in terms of its projection to our human species is that it seems that despite our presumed evolution, we are still expressing ancestral animal drives camouflaged by cultural means.

Thus, despite what Van Horik et al. (2012) stated, that increased cognitive capacity:

> ... is thought to have arisen independently across distantly related species through processes of convergent evolution, driven by the need to solve comparable social and ecological problems, and ... evolutionary divergence has resulted in the independent evolution of different neuroanatomical components and structures across species.

Our species has not only proved to be incapable of overcoming this ancestral animal drive (combination of dominance/prevalence and unequal rights). Additionally, it applied its cognitive capacity to develop cultural and institutional means to cramp this animal drive to social structures, also, to impose it either financially, military, or by subtle or overt psychological media means. This action attempts to camouflage or obscure the true power domain and the generation of inequalities that deeply affect human development. Neural circuits subserving evolutive ancient, basic, animal drives reside below paleo- and neocortical levels, within basal brain structures. Thus, on metaphorical grounds, more recent brain developments took place on an already occupied basal, ground-building neural circuits. These circuits condition our social behavior based on primeval physical needs and emotional settings. We can build on top of it the most daring and beautiful behavioral expressions and goals or tint them with ancient, basal drives.

The above statement on our behavioral domain, coupled with technological development at various levels of human life that painstakingly unequally reaches ordinary people, provides a false impression: that evolution in the collective quality of life represents a universally shared condition. Additionally, the illusion that – in general– for political and financial leaders, the latter represents a primary concern that stands before personal or corporative profit or other life quality and technological privileges. Some of them were mentioned earlier (Colombo 2020, 2021a, b) and will be briefly mentioned below.

This bleak picture is complicated further by the ancient tribal-like type of competition for territories, ideologies, and religious beliefs, whether earthly or in outer space, attracted by economic priorities and subsequent power gains. That is, dominance/prevalence (whether financial, military, gregarious, or religious) has cracked the badly needed utopic vision of humankind evolving despite its cultural differences, i.e., in terms of conceding priority to common needs for human development. This would give support to the early statement by Hamilton (1963):

> It follows that altruistic behavior which benefits neighbors irrespective of relationship (such as the warning cries of birds) will only arise when (a) the risk or disadvantage involved is very slight, and (b) the average neighbor is not too distantly related.

Population Growth and Unequal Rights: Grounds for Overt and Hidden Dominance/Prevalence

According to Powell et al. (2009):

> … demography is a major determinant in the maintenance of cultural complexity and that variation in regional subpopulation density and/or migratory activity results in spatial structuring of cultural skill accumulation.

Also quoted in d'Errico and Stringer (2011):

> Powell et al. [9] reach the conclusion that the number and size of subpopulations and the degree of interaction between them are key factors in the emergence, maintenance, spread and loss of innovations. They speculate that population size in Africa could have reached a critical threshold about 100 000 years ago, when population density and enhanced contact between groups could have allowed the rate of accumulation of innovations to significantly overtake their loss.

However, there may be a point within the population growth curve in which socioeconomic stratification hampers the said effect of critical population size. This may compound the emergence, spread, and loss of innovations, limiting it to favored subpopulations. Stratification of cognitive awareness and cultural interests would emerge from socioeconomic disparities and public media management of subpopulations' interests and concerns by corporations' vested interests, as it has been currently shown. The crackling of community cultural development and level of public affairs awareness is expressed in open or hidden autocratic societies, the latter as in pseudo-democratic ones where public interests are molded to fit corporation—political, religious, or financial—profit or power advantage goals.

Following Foley and Lahr (2011), cultures result from group fissions that depend on kin-based communities' reproduction over generations, affected by ecological and geographical factors and socially transmissible behaviors. According to these authors, the latter would structure the outcome of the said group fission that will define group boundaries, and such boundaries would represent a significant factor in the evolutionary ecology of culture.

In modern times, these boundaries are affected by cultural and material production profiles and their competitiveness in regional and globalized markets. When such production profiles break through local standards and enter globalized markets, their competitive profit goals become multinational. In this corporative arena, promotion profiles and competitive strategies to maximize profit replace the idea of "human beings" by "potential consumers," i.e., anomic numbers in a corporative world of profit-seekers. People become a pawn in a chess game where their only chance to acquire social identity is to seek collective actions.

As in any network information, transfer and access become crucial at biological and social domains:

> A central idea in contemporary biology is that of information. Developmental biology can be seen as the study of how information in the genome is translated into adult structure, and evolutionary biology of how the information came to be there in the first place. (Szathmäry and Smith 1995)

In a modern world where information access and implementation define the probability of individual and collective development and well-being, human's social life (dominated by different degrees of power prevalence and survival needs) is flooded with Machiavellian behaviors that excel in general politics, as reflected in the following profile:

> According to Machiavelli, a ruler with a clear agenda should be open to any and all effective tactics, including manipulative interpersonal strategies such as flattery and lying. (Jones and Paulhus 2009)

On individual developmental grounds, the impact of newborn physical and emotional raising conditions on individual and collective construction has been reviewed by several authors at the experimental and field domains. This was considered in previous chapters and made explicit also by Uller (2012):

> ... mothers and fathers have a significant impact on their offspring well beyond birth in many species through resource provisioning and behavioural interactions, which play important roles in the ontogeny of species-typical phenotypes and individual phenotypic variation (Gottlieb 1997; Avital and Jablonka 2000; Maestripieri 2009; Michel 2011).

The Warmonger Human: Technology Development and Building of Power; When Dominant/Prevalence Behavior Merges into Military Technology Development

The ancestral drive toward dominance and conflictive encounters has powered technological development toward the probability of warfare events. Financially, this development took place at the expense of maximizing population welfare development and worldwide life quality continuous improvement, with the additional promotion of political corruption levels (Colombo 2015).

According to the Carnegie Endowment for International Peace (Schoff and Ito 2019):

> Some international relations analysts and historians point out that AI technology could bring about a 'Second Great Divergence' of productivity –allowing countries and firms that are the earliest and most successful adopters to leap ahead of other peers– following the First Great Divergence brought about by the Industrial Revolution. Such technological innovation is affecting virtually all fields.

According to the New Scientist[2], the Defense Advanced Research Projects Agency (DARPA) (USA) plans to put a nuclear-powered rocket in orbit (Fig. 10.1) by 2025:

> The project will focus on satellites in orbits up to 400,000 kilometers, which is far higher than current space operations carried out by military spacecraft... With the technology,

[2] https://www.newscientist.com/article/2274199-the-us-plans-to-put-a-nuclear-powered-rocket-in-orbit-by-2025/#ixzz6s2Y2GNC5

> military operators could move nuclear-powered communications or spy satellites at will to an area of interest. It also could open up new possibilities, such as tracking and identifying rival stealth military satellites.

Besides protection strategies against possible meteor threats, extraterrestrial exploration for power and exploitation rights of strategical components[3] and space exploration and colonization represent a scientific investment feast amid earthly bound urgent needs. These needs are due to numerous population inequalities, basic survival and unmet developmental cognitive needs, and ecological degradation.

The following brief account provides evidence that humans have perfected ancient dominance and prevalence animal drives involved with the universals of behavior and became highly sophisticated, specialist, warmongers. The following examples illustrate open and covert trends regarding power (military) readiness in prevalent industrialized nations.

Fig. 10.1 "The US plans to put a nuclear-powered rocket in orbit by 2025. This would make it ideal for prolonged maneuvering in orbit." (New Scientist, SPACE 12 April 2021)) Nuclear thermal rockets could be more maneuverable in space. (DARPA)

[3] https://www.nationalgeographic.com/science/article/spacex-launch-kicks-off-regular-commercial-flights-into-orbit?cmpid=org=ngp::mc=crm-email::src=ngp::cmp=editorial::add=Science_20210421&rid=95527DDF86250D8AB073474627963407

China[4]

- The Super Great White Shark stealth helicopter
- DF-41 missile:

  Much of the parade hype has focused on this powerful intercontinental-range ballistic missile, thought to be the mainstay of the People's Liberation Army Rocket Forces (PLARF) arsenal for years to come -and, by some estimations, the most powerful missile on the planet.

- JL-2 submarine-launched ballistic missile (SLBM):

  Each sub can carry 12 of the single-warhead JL-2 missiles. With an estimated range of 4,473 miles (7,200 kilometers), it is regarded as more of a regional than global weapon.

- DF-17:

  China has been testing HGV (**hypersonic glide vehicle**) technology since 2014 and is expected to deploy it in 2020, according to the Missile Defense Project. The DF-17 will be **capable of carrying both nuclear and conventional warheads**, it added.
  Additional developments: DR-8 stealthy drone, Drone submarine, Sharp Sword drone.

USA

- Naval Strike Missile:

  US Navy adds powerful new missile in Pacific expected to have an operational weapon before 2022. The USS Gabrielle Giffords, a sleek, speedy, low-profile littoral combat ship, left San Diego earlier this month carrying the US Navy's new Naval Strike Missile and a drone helicopter that helps aim it.[5]

- Tactical Ultrashort Pulsed Laser for Army Platforms:

The US Army is building the most powerful laser weapon in the world (2019)[6] (Fig. 10.2).

Additional examples that the world is sitting on a bed of hot embers fed by competing for territorial, space, and financial dominance and survival are as follows:

- MQ-9 Reaper military drone:

  The launch on the mightiest rocket built by the United Launch Alliance (ULA): the massive Delta IV Heavy, carrying the classified NROL-82 **spy** satellite into orbit for the National Reconnaissance Office. AI is helping US Air Force to decide which targets to strike. The

[4] https://edition.cnn.com/2019/09/27/asia/china-military-parade-analysis-preview-intl-hnk/index.html

[5] https://edition.cnn.com/2019/09/10/asia/us-navy-naval-strike-missile-asia-pacific-intl-hnk/index.html

[6] https://www.newscientist.com/article/2268553-the-us-army-is-building-the-most-powerful-laser-weapon-in-the-world/#ixzz6n5FyBmZW

Fig. 10.2 Tactical Ultrashort Pulsed Laser for Army Platforms. (John F. Williams/U.S. Navy) (Ibid)

Secretary of the Air Force Frank Kendall said that the US Air Force (USAF) had recently deployed AI algorithms for the first time to a live operational kill chain[7] (Fig. 10.3).

DEATH BY ALGORITHM

The US military has started using AI to guide its air strikes, according to Secretary of the Air Force Frank Kendall. Speaking at the Air Force Association's Air, Space and Cyber Conference in National Harbor, Maryland, on 20 September, Kendall said that the US Air Force had recently 'deployed AI algorithms for the first time to a live operational kill chain'. The kill chain is the process of gathering intelligence, analysing, and evaluating it and directing weapons to destroy a target.

Russia[8]

- TK-17 Arkhangelsk:
- It is a Project 941 Akula nuclear-powered ballistic missile submarine[9] (Fig. 10.4).

[7] https://www.newscientist.com/article/2291586-ai-is-helping-us-air-force-to-decide-which-targets-to-strike/#ixzz77z1Knp2p

[8] https://tradingeconomics.com/russia/military-expenditure

[9] https://www.world-of-engineering.org

Fig. 10.3 A US MQ-9 Reaper military drone. USA (Department of Defense Archive/Alamy)

Military expenditure in Russia is expected to reach USD 70 billion by the end of 2021, according to Trading Economics' global macro models and analysts' expectations (CNN).[10]

> Russia is amassing unprecedented military might in the Arctic and **testing its newest weapons** in a region freshly ice-free due to the climate emergency, in a bid to secure its northern coast and open up a key shipping route from Asia to Europe. The head of Norwegian intelligence, Vice Admiral Nils Andreas Stensønes, told CNN that his agency has assessed the Poseidon as 'part of the new type of nuclear deterrent weapons. And it is in a testing phase. But **it's a strategic system, and it's aimed at targets ... and has an influence far beyond the region in which they test it currently**.
>
> This unmanned stealth torpedo is powered by a nuclear reactor and intended by Russian designers to sneak past coastal defenses – like those of the US – on the seafloor. **The device is intended to deliver a warhead of multiple megatons, according to Russian officials, causing radioactive waves that would render swathes of the target coastline uninhabitable for decades. In November, Christopher A Ford, then assistant secretary of state for International Security and Non-Proliferation, said the Poseidon is designed to inundate U.S. coastal cities with radioactive tsunamis**. (Bold characters by author J.A.C.)

To underline present tendencies in the development of war machines, besides orbital technology, the US Army is developing the Tactical Ultrashort Pulsed Laser for Army Platforms, a laser weapon supposedly more than a million times more

[10] https://tradingeconomics.com/russia/military-expenditure

Fig. 10.4 TK-17 Arkhangelsk, a nuclear-powered ballistic missile submarine (https://www.world-of-engineering.org)

powerful than existing ones: the ultrashort pulse would vaporize the surface of a target rather than melt it, as informed by the New Scientist[11].

With the uncrewed spacecraft launched to Mars by the United Arab Emirates, there will be seven orbiters over that planet[12]. Thus, the race is on, and the extraterrestrial space is up for grabs and competitive development. The unsolved conflicts among nations and corporations are mounting an extraterrestrial stage.

[11] https://www.newscientist.com/article/2268553-the-us-army-is-building-the-most-powerful-laser-weapon-in-the-world/?utm_source=nsday&utm_medium=email&utm_campaign=NSDAY_220221

[12] https://www.newscientist.com/article/2266796-the-united-arab-emirates-hope-orbiter-isabout-to-arrive-at-mars/?utm_source=nsday&utm_medium=email&utm_campaign=NSDAY_080221

In present times, do these multimillionaire projects represent an expression of their social indifference—of their political and corporative-based power/profit-seeker sociopath behavior—toward urgent, priority needs of fellow citizens? This diverts required funds and strategies away from the population's well-being and educational/cognitive development.

Really, how far are we from the following comparison with chimpanzees tested in the ultimatum game analyzed by Jensen et al. (2007)?

> …we show that in an ultimatum game, humans' closest living relatives, chimpanzees (*Pan troglodytes*), are rational maximizers and are not sensitive to fairness. These results support the hypothesis that other regarding preferences and aversion to inequitable outcomes, which play key roles in human social organization, distinguish us from our closest living relatives. It thus would seem that in this context, one of humans' closest living relatives behaves according to traditional economic models of self-interest, unlike humans, and that this species does not share the human sensitivity to fairness.

Regarding competitive power management, we have not succeeded in getting over the deeply anchored ancestral animal drives based on dominance/prevalence. The promoted human sensitivity to fairness is at stake and limited to special situations (major crisis) and independent action groups or individuals.

Fear Conditioning and Degrees of Freedom

The concept of profit in human civilization culture is intimately bound to ancient drives that fueled strategies for survival and privileged feeding and reproductive standings. Nowadays it is directed toward dominance/prevalence goals projected to strategies of corporative organizations, political prevalence, or social class-bound privileges. The question then arises when progress implies unbalanced corporative competition in which lead is obtained due to market advantage position, beyond ordinary people's awareness and control. In such a case, how could ancient drivers of competition aimed at maximizing survival probability and profit (with diverse practices of deceiving, consumer lying, unethical marketing practices, and dominant position) be eliminated from the collective idea of community progress and security? It, thus, sounds as non-real the following comparative statement made by Jensen et al. (2007), given the manifest behavioral bias and use of power aimed at self- or group-interest profit and power gain in our species:

> It thus would seem that in this context, one of humans' closest living relatives behaves according to traditional economic models of self-interest, unlike humans, and that this species does not share the human sensitivity to fairness.

Rather, considering present conditions and trends relative to social and cognitive opportunities and inequalities and covert public opinion controlled by media and marketing strategies would suggest the opposite be true.

Compounding this picture is the general lack of proper access to information and cognitive means to process it. This generates profound gaps among social strata, which potentiate inequalities at all levels of social life, most clearly, generating quasi-stable conditions for the generation of dominant and dominated groups.

This basic social condition is not exclusive of our species but also expressed across the natural kingdom. Granted, human cultural development has generated sophisticated institutionalized forms of dominance/prevalence and of social resistance behaviors (whether politically or ideologically organized, or as sociopathic behaviors). So far, the concepts of equal rights and opportunities represent a utopic horizon, a declaration of unpracticable principles under current conditions. So, one question arises regarding how to modify what appears to be a natural expression in large groups of social animal life, perhaps with counted exceptions in some comparatively small, isolated, ancient human communities. Furthermore, if not possible, how to construct societies in which labor or occupational strata do not derive into overt or subtle forms of dominance or oppression?

The impact of population size has been considered, among others, by Knauft (1991) and Lagergraber et al. (2017), respectively:

> Comparative analyses indicating that cooperative territoriality commonly occurs in primate species who live in small groups (5) and field studies demonstrating that individual participation in intergroup encounters decreases with group size (11, 12, 23) have been interpreted as evidence of the difficulty of achieving collective action in large groups. Nevertheless, how some species, including humans, regularly engage in collective action despite living in large groups remains unexplained.
>
> … group augmentation theory helps explain one of the most striking examples of collective action in nonhuman primates, territorial boundary patrolling by male chimpanzees: proposes that individuals should bear the short-term costs of collective action even when they have little to gain immediately if such action leads to increases in group size and long-term increases in reproductive success.

As it happens among human beings, deception and cheating in nonhuman primates also give them reproductive and hierarchical advantages. So, what cannot be met or accomplished by reason or force and fear induction will come by deception or cheating. According to Hauser (1992), from field studies in rhesus monkeys (Macaca mulatta):

> … there are significant costs to withholding information. Such costs may constrain the frequency with which deception occurs in this and other populations.

Penalties for breaches in human social life are institutionally penalized, with religious traditions acquiring various levels of severity throughout their history, as influenced by dogmas based on supernatural beliefs and hierarchical status. Social standing (emanating from political, religious, social, or financial quarters) introduces conditioned penalties accordingly. Hence, besides economic, equal rights, and social strata, an additional one ensues because of them, i.e., how to enforce responsibility and penalty applied with equal weight among the various actors within socioeconomic strata. Repressive actions or coercion to induce fear are widespread among social animals to enforce hierarchies, feeding, and reproductive rights, whether implemented by individual leadership or by selected or appointed (genetically or behaviorally formalized) kin groups (or prevailing social classes). Such behaviors have been either open or culturally disguised and institutionalized in most human societies across their history.

Given evolutive inertias and self-convenient policies, a basic cultural change must take place to override such determinant trends. They condition individual and collective degrees of freedom—within the current, dynamic limitations of our neurobiological construction (Colombo 2014).

Space Exploration, Power Competition, and Colonization

> On 27 January 1967, the United Nations General Assembly signed the 'Treaty on the Principles' that should govern activities of States in the exploration and use of outer space, including the Moon and other celestial bodies, known as the Space Treaty. Article 2 states that Outer space, even the Moon and other celestial bodies, may not be nationally appropriated by claiming sovereignty, use or occupation, or in any other way.[13]

> According to NASA, there are currently millions of pieces of space junk in the range of altitudes from 200 to 2,000 kilometers above the Earth's surface, which is known as low Earth orbit (LEO). Most of the junk is comprised of objects created by humans, like pieces of old spacecraft or defunct satellites. This space debris can reach speeds of up to 18,000 miles per hour, posing a major danger to the 2,612 satellites that currently operate at LEO.[14]

Private entrepreneurs, such as SpaceX, hinted what might develop soon since they appear to override previous international principles on outer space. Instead of promoting international consensus on extraterrestrial exploration, SpaceX declares that it will follow the law as governed by the state of California in the United States and for services provided on Mars or in transit via starship or other colonization spacecraft. Thus, it would thus seem that it will be up for grabs[15, 16]:

> … the parties recognize Mars as a free planet and that no Earth-based government has authority or sovereignty over Martian activities. Accordingly, Disputes will be settled through self-governing principles, established in good faith, at the time of Martian settlement.

The Carnegie Endowment for International Peace has emitted a balance of economic and military power, prompting governments and large corporations to compete aggressively now over their development and applications.

[13] Translated from https://www.infobae.com/sociedad/2019/07/12/el-dueno-de-la-luna-entrevista-con-el-estadounidense-que-vendio-terrenos-a-mas-de-6-millones-de-personas/

[14] https://www.sciencedaily.com/releases/2021/02/210227083314.htm

[15] https://futurism.com/the-byte/starlink-terms-of-service-demands-users-recognise-mars-free-planet

[16] https://www.inverse.com/innovation/spacex-mars-city-starlink-terms-of-service-declares-free-planet

This calls for concern in China's potential growth on technological grounds:

> Aside from broader issues of trade and economics, the United States and Japan should consider the specific risks and opportunities related to competition with China in high-tech innovation. A so-called fourth industrial revolution is under way, a revolution characterized by discontinuous technological development in areas like artificial intelligence (AI), big data, fifth-generation telecommunications networking (5G), nanotechnology and biotechnology, robotics, the Internet of Things (IoT), and quantum computing. Breakthroughs in these fields can potentially shift the future (Schoff and Ito 2019).

Furthermore, besides corporate revenue interest, military objectives have always been involved. They are implicit in the announcement of the US Department of Defense stating that Elon Musk's company and United Launch Alliance (ULA), a joint project of Boeing and Lockheed Martin, would share the job between 2022 and 2026 in launching into space the military's highest-priority satellites. This "national security" cargo would include intelligence gathering, GPS, and military communications[17]. This stems from historical grounds with objectives shared by other military and financial powers, a race that places the citizens of our world as astonished spectators unless human values and objectives can make a sharp turn.

Thus, politics and strategies involved in Earth-bound and extraterrestrial-bound explorations are evolving under explicit and hidden regulations that forecast a projection of past misadventures at high human and financial costs. This development appears driven as deeply entrenched with ancient animal drives that ruled animal evolution on planet Earth. Human culture and ideologies attempt to mask real intentions of dominance and prevail in these domains. Changes in this behavioral profile must occur during unforeseeable lengths of time before the species can progressively change ancient drives by mutation, replacement, or cultural conditioning of ancient neural circuitries and behavioral drives entrenched in our developmental animal history.

Several technological advances (e.g., nuclear atom fission and fusion) and extraterrestrial projects had deceitful arguments and consequences. Undoubtedly, other developments have profound consequences on our knowledge about our human dimension and potential quality of life. However, human knowledge usually has a bifrontal face, for its ancient animal drives of dominance and prevail come hand in hand with the former. They express strategic power gains in the game of maximizing dominance and profit. Many technological advances (e.g., nuclear atom fission and fusion) and extraterrestrial projects had deceitful arguments and consequences. Undoubtedly, other developments have profound consequences on our knowledge about our human dimension and potential quality of life. The following argument sounds naive in this context and hides the other side of the coin, the one that is bound to our ancestral animal drive as a species:

> Plenty of ink has been spilled over the environmental impacts of space travel. It's costly, it can be wasteful, and debate swirls over the potential carbon emissions. But to balance the

[17] https://www.livescience.com/spacex-ula-win-pentagon-rocket-contratc/html?

> scales, space travel arguably helped launch the environmental movement by clearly showing us the fragility of our home planet, and satellite data plays a huge role in our understanding of modern climate change.[18]

The few briefs mentioned above are a pale example of power deployment on Earth and beyond. There is a comparatively feeble search for equilibrium among war-leading economies. With military power concentrated in a few nations, collective and individual freedom appears an unreal and biased concept for the world and defines its future as a fragile construction. Relative power strength implies a strategic standing within and between national affairs.

There is a profound comparative gap with a group of nations and corporate organizations regarding rights to apply expenditures to sustain an unsettling career on power dominance (financial or military) or attempting to reach a stale state, biased by the need to equalize or outcompete the opponent. This has generated a profound gap between the notion of the scientific pursuit of knowledge and collective gain in social improvement and power-seeker groups at the political and corporate levels. The critical consequence is that most world citizens that primarily act as a stratum of spectators are amazed at the small aliquots of information and decision they have access to, on a pile of unreachable piled up strata of information access, as described in Colombo (2020).

If we disregard the illusion of true democracy (which entails full information and the capability to access and decode it), the state of the world will closely resemble eusocial species. That is a decision-making caste-like social organization formed by an upper stratum of multibillionaires, corporations, and political leaders. Below them, strata of laborers work to support their doings and either perform the needed consumerist behavior or contribute to expand, build, and deploy war machinery.

References

Boesch C. Joint cooperative hunting among wild chimpanzees: taking natural observations seriously. Behav Brain Sci. 2005;28:692–3.

Boly M, Seth AK, Wilke M, Ingmundson P, Baars B, Laureys S, et al. Consciousness in humans and non-human animals: recent advances and future directions. Front Psychol. 2013;4:1–20. https://doi.org/10.3389/fpsyg.2013.00625.

Clutton-Brock T. The evolution of society. Phil Trans R Soc B. 2009;364:3127–33.

Colombo JA. "De la nature humaine", Réfractions. 2014;127–134. Paris. https://doi.org/10.1098/rstb.2009.0207I. ISBN 978-2-918697-09-1.

Colombo JA. *Los Homo sabios* (The wise Homo), Ed. Buenos Aires Books, Buenos Aires; 2015.

Colombo JA. *Creativity, a profile for or species*. Newcastle: Cambridge Scholars Publishing; 2020.

[18] https://carnegieendowment.org/2019/10/10/competing-with-china-on-technology-and-innovation-pub-80010

Colombo JA. *The Homo within the sapiens*. New York: Nova Science Publishers; 2021a.
Colombo JA. Ancestral drives and the complex behavior of our species. Int J Primatol Res. 2021b;19:18–9.
D'Errico F, Stringer SB. Evolution, revolution, or saltation scenario for the emergence of modern cultures? Phil Trans Roy Soc B. 2011;366:1060–9.
Dawkins R. *The selfish gene*. Oxford: Oxford University Press; 1995.
Foley RA, Lahr M. The evolution of the diversity of cultures. Phil Trans Roy Soc B. 2011;366: 1080–9.
Gardner RA. Animal cognition meets evo-devo. Behav Brain Sci. 2005;28:699–700. https://doi.org/10.1017/S0140525X05490123.
Hamilton WD. The evolution of Altruistic behavior. Am Nat. 1963;97:354–6.
Hauser MD. Costs of deception: cheaters are punished in rhesus monkeys (Macaca mulatta). Proc Natl Acad Sci USA. 1992;89:12137–1213. https://doi.org/10.1073/pnas.89.24.12137.
Jensen K, Call J, Tomasello M. Chimpanzees are rational maximizers in an ultimatum game. Science. 2007;318:107–9.
Jones DN, Paulhus DL. Chapter 7: Machiavellianism. In: Leary MR, Hoyle RH, editors. Individual differences in social behavior; 2009. p. 93–108.
Knauft B. Violence and sociality in human evolution. Curr Anthropol. 1991;32(4):1983.
Lagergraber KE, Watts DP, Vigilant L, Mitani JC. Group augmentation, collective action, and territorial boundary patrols by male chimpanzees. PNAS. 2017;114:7337–42.
Mao Y, Catacchio CR, Hiller LW, Porubsky D, Eichler EE. A high-quality bonobo genome refines the analysis of hominid evolution. Nature. 2021;59:477–81.
Nottebohm F, Nottebohm ME, Crane L. Developmental and seasonal changes in canary song and their relation to changes in the anatomy of song-control nuclei. Behav Neural Biol. 1986;46:445–71. https://doi.org/10.1016/s0163-1047(86)90485-1.
Penick CA, Ghaninia M, Haight KL, Opachaloemphan C, Yan H, Reinberg D, Liebig J. Reversible plasticity in brain size, behaviour and physiology characterizes caste transitions in a socially flexible ant (*Harpegnathos saltator*). Proc R Soc B. 2021;288:20210141. https://doi.org/10.1098/rspb.2021.0141.
Powell A, Shennan S, Thomas M. Late Pleistocene demography, and appearance of modern human behavior. Science. 2009;324:1298–301.
Richerson PJ, Boyd R. *Not by genes alone. How culture transformed human evolution*. Chicago/London: The University Chicago Press; 2005.
Robertson DR. Social control of sex reversal in a coral-reef fish. Science. 1972;177:1007–9. https://doi.org/10.1126/science.177.4053.1007.
Sapolsky RM. Social status and health in humans and other animals. Annu Rev Anthropol. 2004;33:393–418.
Sapolsky RM. The influence of social hierarchy on primate health. Science. 2005;308:648–52. https://doi.org/10.1126/science.1106477.
Schoff JL, Ito A. *Competing with China on Technology and Innovation*. Washington, DC: Carnegie Endowment for International Peace; 2019.
Schuster R. Why not chimpanzees, lions, and hyenas too? Behav Brain Sci. 2005;28:716–7.
Szathmäry E, Smith JM. The major evolutionary transitions. Nature. 1995;374:227–33.
Tomasello M, Carpenter M, Call J, Behne T, Moll H. Understanding and sharing intentions: the origins of cultural cognition. Behav Brain Sci. 2005;28:675–735.
Tramontin AD, Brenowitz EA. Seasonal plasticity in the adult brain. TINS. 2000;23:251–60.
Uller T. Parental effects in development and evolution. In: Royle NJ, Smiseth PT, Kölliker M, editors. The Evolution of Parental Care. 1st ed. Oxford University Press; 2012.
van Horik JO, Clayton NS, Emery NJ. Chapter 5: Convergent evolution of cognition in Corvids, Apes and other animals. In: *The oxford handbook of comparative evolution-*

ary psychology. New York: Oxford University Press; 2012. https://doi.org/10.1093/oxfordhb/9780199738182.013.000.
Whitten A, Byrne RW. Machiavellian intelligence II. Extensions and evaluations. Cambridge, UK: Cambridge University Press; 1997.
Zink CF, Yu Tong Y, Chen Q, Bassett DS, Stein JL, Meyer-Lindenberg A. Know your place: neural processing of social hierarchy in humans. Neuron. 2008;58:273–83.

Chapter 11
Factors of Our Human Future

> ...it seems that a bipolar condition has dominated our human dimension as it has led to astounding technological developments in the face of unacceptable humane conditions, affecting almost one-third of our global population, with wide segments of it seemingly unable to integrate into the evolving technological (technocratic?) social dimensions and related scientific fields. It is a knowledge gap that seems to be widening with each new development and with time. The optimistic view that such development will overflow into all social levels has been crushed by history and the present state of the world so far. The historical gap among social classes and ethnicities seems to be now potentiated by mandatory fast-tracked development and training on cognitive and informational grounds. (Colombo 2021)

Previous chapters approached the analysis of fundamental evolutionary ingredients of animal behavior and some of the consequences at the level of our human doings. They are considered universal components of the natural kingdom, masked by ecological and cultural adaptations, and inserted as the primary platform of species variants regarding behavioral and physical phenotypes.

The crackled planetary view involves socially, culturally, religiously, and ecologically human variants acting on neurobiological genetic inheritance. This suggests different slopes of communitarian development, goals, and profiles, besides shared (globalized) components, involving technological and financial domains. Hence, the future of human society cannot be placed under a single, simple equation but rather augurs continued, diverse evolutive routes based on inequalities and evolutionary cultural dystopias on cognitive, emotional, and economic grounds.

In this multifactorial puzzle, perhaps technology, religiosity, and enduring social inequalities appear as principal factors. (See metaphorical representation of *tectonic plate friction* in Fig. 11.1.)

> While creativity provides the main constructive profile of our species, the primal, ancient, animal drive for dominance places its future at risk. This combination of bipolar natures generates a distorting impact on the global perspective of our collective future and ecological conditions. Therefore, those (creative) minority groups that can embrace the most daunting scientific-technological challenges productively and fight off the avarice trends of elitist groups leave marginalized individuals on a different social evolutive path. This is

J. A. Colombo, *Dominance Behavior*, https://doi.org/10.1007/978-3-030-97401-5_11

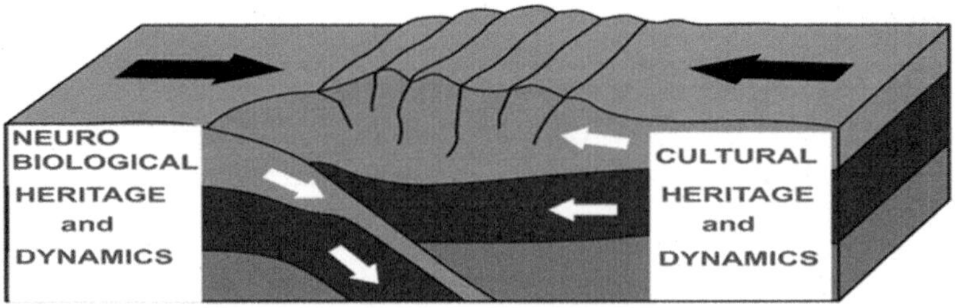

Fig. 11.1 Metaphorical representation of the "neurobiological" and "cultural" "tectonic plate friction," as a source of adapted/nonadapted/pathological behaviors. (Colombo 2019)

> either due to a lack of accessibility to progress or different cultural values and priorities. (Colombo 2019)

Fear is one fundamental driver in animal behavior, including humans. According to Montardy et al. (2021), fear of predation in the rodent and marmoset monkey activates the medial hypothalamic brain stem defensive system, and it is highly conserved on evolutive grounds. In humans, electrical stimulation of the ventromedial hypothalamus elicits feelings of intense fear capable of triggering panic attacks (Wilent et al. 2010). Thus, physical and virtual (cultural) menaces can induce a highly conserved visceral-somatic response of fear in the recipient.

> Our findings demonstrate that the ventromedial hypothalamus is recruited by predator exposure in primates and that anatomical connectivity of the rodent and primate medial hypothalamic defensive system is highly conserved. (Montardy et al. 2021)

Within human cultural domains, fear—whether through manifest physical censorship or induced by emotional means—modulates several of our individual and collective behaviors. Thus, the notion arises that the virtual superstructure (beliefs) with which we humans construct behavioral and cultural webs has underlying connections with an ancestral fear web of an ecological and biological nature. This highly conserved universal means by which individual and collective behaviors can be mastered represents a fundamental means to induce acceptance or passivity toward political and religious structures and advances. At the human level, the concept of predation represents an equivalent dimension to intellectual and behavioral censorship and dominance by factual or imaginary (virtual) authorities.

Critical Elements in Our Cultural Puzzle: Rituals, Social Connectivity, and Coherence; Rituals and Moralizing Gods

> Anthropologists and sociologists have speculated that rituals involving synchronous activity may produce positive emotions that weaken the psychological boundaries between the self and the group... endowing some cultural groups with an advantage in societal evolution. (Wiltermuth and Heath 2009)

According to Boyer and Lienard (2006), ritualized behaviors would be secondary products of cognitive architecture. With gestural redundancy and repetitive, predictable movements, such behavior would tend to confer a sense of control and prediction of anxiety- or fear-generating events for alleged potential threats. Regardless of its structural complexity, as expressed in human rituals in different cultures, according to Boyer and Lienard (2006), the ritualistic behavior is the expression of a *hazard precaution system* directed toward the detection and reaction to *supposed risks*—as different from *actual risks* that induce flight behavior.

As previously suggested (Colombo 2019), during ancient times of the genre *Homo*, the need to build up a "manageable" relationship with natural forces that are often overwhelming, and to glue intragroup relationships, probably triggered a web of cognitive and emotional behaviors. Attribution of intentionality and agency to various objects and animals, as well as to "low" and "high" gods, must have eased up those feelings and the possibility of "making deals" within their earthly, intimal realm. Botero et al. (2014) conclude that certain beliefs in "moralizing gods" tend to promote cooperation when more prominently poor or deprived is the environment and more significant the risk of exposure to environmental stress. Given the strong correlation observed between cooperative behavior and ecological uncertainty, Botero et al. (2014) conclude that their observations support the notion of an existing link between ecology and religion (or beliefs in general):

> Given the strong correlation between cooperation and ecological uncertainty in nonhuman animals (2, 3), these findings are especially suggestive of a link between ecology and religion.

Mithen (1996) underlines two crucial periods in cultural development:

> The two really dramatic transformations in human behavior occurred long after the modern size of the brain had evolved. They are both associated exclusively with Homo sapiens sapiens. The first was a cultural explosion between 60,000 and 30,000 years ago when the first art, complex technology, and religion appeared. The second was the rise of farming 10,000 years ago. (Mithen, 1996)

The ability to transfer emotional states generated by fear, or fear to natural events of a significant magnitude or unexpected nature, to virtual—fantastic—cognitive constructions ended up conditioning individual and community thoughts and behavior. Conditioning aimed to avoid the supposed threat of uncontrollable forces assigned power and terrestrial designs to agents or moralizing gods.

As Peoples and Marlowe (2012) suggest:

> ...that belief in moral High Gods was fostered by emerging leaders in societies dependent on resources that were difficult to manage and defend without group cooperation. These leaders used the concept of a supernatural moral enforcer to manipulate others into cooperating, which resulted in greater productivity.

Extreme rituals amplify prosocial attitudes and behaviors. Those events also reinforce group ideologies from which it is intended to exercise social dominance (Sidanius et al. 2004):

> …the acceptance of ideologies that legitimize inequality and behaviors that produce inequality is partly determined by people's general desire for group-based dominance. This desire for group-based dominance is captured by a construct we call social dominance orientation.

Inzlicht et al. (2012), like other authors as Barrett (2011), speculate about the adaptive value of religious thought and rituals:

> This model holds that people are motivated to create and sustain meaning (i.e., a sense of coherency between beliefs, goals, and perceptions of the environment, which provides individuals with the feeling that the world is an orderly place), and that religious beliefs buffer the distress associated with disruptions to meaning, thus leading to decreases in distress. (Inzlicht y cols., 2012)

This interpretation would provide an adaptive vision fundamentally in times of crisis, whether of collective or individual nature.

As mentioned in Colombo (2019):

> … The need to build conceptual structures that shield people from natural threats and death, generated gods, rituals, idols, instruments – amulets –, compulsive behaviors, pilgrimages, sacrifices of different sizes with the ability to pretend to rule or correct behavior or 'destiny'. From that imaginary world, from that emotional submission – later elaborated in a cognitive form– arose rules, ethical norms, commandments, institutions that regulated the admitted behavior. This mystical relationship with the unknown and fear-threatening agents provided a resource to modulate, regulate or tame societies that became progressively more complex…

In this regard, development of conceptual structures supporting cultural complexities has been linked to population size, not without opposing views. As summarized in Powell et al. (2009):

> Demographic factors can thus explain geographic variation in the timing of the first appearance of modern behavior without invoking increased cognitive capacity.

And, furthermore:

> UP (*Upper Paleolithic*) material culture, usually referred to as the Late Stone Age (LSA) in Africa, is characterized by a substantial increase in technological and cultural complexity, including the first consistent presence of symbolic behavior, such as abstract and realistic art and body decoration… (Italics by JAC)

In the authors words:

> This presupposes the requisite of biologically determined cognitive capacities from the time of origin some 160 to 200 ka.

This linking of developmental conceptual structures to population size has been supported by Henrich et al. (2016) and questioned by Vaesen et al. (2016) based on archeological records.

The Educational Conundrum: Progressive Technological Dependence, a Route Toward Freedom or Lack of It?

There is this disturbing question of Harari (2015):

> When bodies and brains become designer products, will natural selection give way to (technological) intelligent design?

For, who will be the "master designer" or what public cultural awareness level will allow such type of decisions? As stated earlier (Colombo 2007, 2019):

> Educational public policies that advocate knowledge, the exercise of imagination and divergent thinking or conceptual expansion can only be effective if they are integrated into equity in social construction and the optimization of individual freedom. Indigence, poverty, privilege and class prejudice are its powerful opponents.
>
> …post-industrial societies are becoming increasingly dependent on material and technological culture to the point of 'embraining' (sic) them. That is, incorporating both into basic cognitive daily constructions; conceptually becoming *technological hybrids*. Whether this represents a developmentally 'must' or an uncontrolled 'spin-off' of human inventiveness, represents a point that ought to be taken immediate conscience of, at the social and political level.

Besides the improvement in literacy statistics worldwide during the last 200 years or so, the rate of informational change has created a deep gap in terms of amount, quality, and complexity of information and cognitive ability to comprehend and integrate them, not comparable with previous decennials. This produced a cognitive lag that increased with time passed. The following tables intend to approach the actual condition based on only one dimension—the literacy rate among adults.

According to the World Bank (2019),[1] nearly two-thirds of illiterate adults in the world are female. In total, there are about 781 million adults worldwide that cannot read or write. While 31 nations (from 195) reported literacy rates of 99.0% or above, the following nations listed have the lowest literacy rates (see Tables 11.1 and 11.2):

- Guinea-Bissau (30.4%)
- South Sudan (34.5%)
- Mali (35.5%)

Table 11.1 Literacy rate per income rate, adult total (% of people ages 15 and above) (https://worldpopulationreview.com/country-rankings/literacy-rate-by-country) UNESCO Institute for Statistics (uis.unesco.org). Data as of September 2020. *License:* CC BY-4.0

	Most recent year	Most recent value
Low income	2019	61
Lower middle income	2019	76
Middle income	2019	86
Upper middle income	2019	96

[1] https://data.worldbank.org/indicator/SE.ADT.LITR.ZS

Table 11.2 Literacy rate, adult total (% of people ages 15 and above)—low income, per country (https://data.worldbank.org/indicator/SE.ADT.LITR.ZS)

Country	Most recent year	Most recent value
Cote d'Ivoire	2018	47
Afghanistan	2018	43
Angola	2014	66
Burkina Faso	2018	41
Burundi	2017	68
Central African Republic	2018	37
Chad	2016	22
Ethiopia	2017	52
Gambia	2015	51
Guinea	2014	32
Guinea-Bissau	2014	46
Haiti	2016	62
Liberia	2017	48
Mali	2018	35
Mauritania	2017	53
Mozambique	2017	61
Nepal	2018	68
Niger	2018	35
Nigeria	2018	62
Pakistan	2017	59
Sierra Leone	2018	43
South Sudan	2018	35
Sudan	2018	61
Togo	2015	64
Yemen	2004	54

- Central African Republic (37.4%)
- Burkina Faso (41.2%)
- Benin (42.4%)
- Afghanistan (43.0%)

According to the World Bank, world poverty rates (2015)[2] calculated at 3.50 and 5.50 U$S/day (2011 PPP) (source, PovcalNet, World Bank Group) indicate:

World poverty at 3.20 U$S: 26.3% (range, 0.1–66.3)
World poverty at 5.50 U$S: 46.0% (range, 1.5–84.5)

Quality of data collection and limitations are also specifically mentioned in the World Bank 2017 report on Poverty and Shared Prosperity (Chap. 2) and Chap. 3 of

[2] http://iresearch.worldbank.org/PovcalNet/introduction.aspx; https://blogs.worldbank.org/opendata/global-poverty-2015-povcalnet-s-new-estimates-and-improved-documentation; http://povertydata.worldbank.org/poverty/home/; https://data.worldbank.org/indicator/SI.POV.DDAY

the 2018 report on Shared Prosperity: Mixed Progress. This is not a minor issue; the recent United Nations IEAG report "A World That Counts" argues that a lack of data can lead to a "denial of basic rights."

The 2020 World Bank report states:

> About a quarter of the global population is living below the US$3.20 poverty line, and almost half is living below the US$5.50 line, compared with less than a 10th living below US$1.90. These figures translate to 1.8 billion people and 3.3 billion people at the US$3.20 and US$5.50 poverty lines, respectively. The number of people living below US$3.20 today is as high as the number of people in extreme poverty in 1990, the starting point of this analysis, which is perhaps one way to illustrate the scale of the challenge that remains at these higher lines. The number of people living below US$5.50 per person per day has barely declined over the past 25 years.

What Do the Numbers Hide?

The indicators mentioned above involve child development conditions and, hence, condition their probability for future work labor insertion in the "society of knowledge" and to become able to critically handle the informational load in their adult life, thus affecting social construction.

Furthermore, the World Bank (2018) report[3] calls attention to the statistics on poverty when the poverty line is at U$S 5.50:

> The US$5.50 line, reflecting basic needs in upper-middle income countries, presents two distressing findings: (1) almost half the world lives on less than US$5.50 per day, and (2) in the regions of the Middle East and North Africa, South Asia, and Sub-Saharan Africa, despite progress in reducing their poverty rates, more people were living on less than US$5.50 in 2015 than in 1990 due to their growing populations.

According to UNESCO (2019), a comparison among high- and low-income countries with age group students 20–24 years with less than 4 years of schooling shows a significant breach between them.[4] The following Table (11.3) according to UNESCO (Colombo 2021) lists the out-of-school rates for 2017.

Table 11.3 Out-of-school rate for children, adolescents, and youth of primary and secondary school age, according to UNESCO (2017). (Colombo 2021)

Low-income countries	32.65
Lower middle-income countries	21.62
Middle-income countries	16.92
Upper middle-income countries	9.01
High-income countries	3.35

[3] http://www.worldbank.org/en/publication/poverty-and-shared-prosperity

[4] See original analysis in https://www.education-inequalities.org/

As mentioned earlier (Colombo 2019):

> … at the rate of present scientific/technological progress the impact on the comparative development of those individuals imply a heavy functional handicap towards their future.

The samples included from the educational and schooling condition in multiple world regions and nations should call the attention in terms of development and the widening gap in technological development. This statement has at least two interpretations related to the desired developmental route, considering that comparative incomes involving less developed countries could be a way of living in which high technological development is not a desired cultural paradigm. In these communities, the concept of welfare and needed high income to acquire a relative dimension does not represent a shared, significant drive, besides the imposed competitive market conditions. Hence, a cultural consideration should determine whether the impoverishment imposed by corporate dominance over the quality of life violates cultural profiles. Nuances and technological developments attempted to be imposed on societies may not be universally culturally acceptable.

Hence—despite some improvement in specific world regions—published media hypocrisy seems to invade the communication channels, providing a false sense of a worldwide, *globalized*, humanitarian progress. Even having access to any book, newspaper, or electronic notebook does not imply the ability to read its contents, comprehend them, and critically take a stand on what is read, nor does it assure a better community interaction and ecological integration. Precisely, the ecological impact of the human factor remains freed and accumulating somber effects. In this regard, it seems as if our social values, priorities, and premises need to be reviewed and redirect our diverse collective creativity toward earthly issues before planning to "colonize Andromeda," which we ought to do some time after solving them. As stated earlier:

> Highly stratified information access, the arsenal of war, the demands of a runaway economy –with periodic financial crises– and the inability to self-sustain without resorting to policies of 'adjustment' and overexploitation of natural and human resources, while lacking efficient and sustainable ways to dispose of our disposals, constitute a set of strategies that do not build a promising future for world majorities, nor for communities attached to ethnic traditions, nor for the planet. (Colombo 2019)

Besides differing cultural paradigms—as noted above—joint responsibility and shared development do not seem to be within the central social values applied by the financial elites of modern society. The following excerpts are a metaphor of the current state of social affairs:

> …Let's mine asteroids—for science and profit.
>
> …Greed is a powerful motivator to get things done.
>
> …Two events in quick succession have transformed the prospects for commercial space activities: the successful rendezvous last week of California-based SpaceX's privately developed Dragon capsule with the International Space Station; and the bold announcement last month of a new asteroid-mining company, Planetary Resources in Bellevue, Washington, backed by deep-pocketed billionaires. (Elvis, Nature 2012)

> The top percentile of wealth holders now owns just over half of the world's wealth and the richest decile 87.7 percent. (Credit Suisse 2015)[5]

> Since 2010, the wealth of this economic elite has grown by an average of 13% per year; six times faster than the salaries of working people who have barely increased an annual average of 2%. Between March 2016 and March 2017 there was the largest increase in history in the number of people whose fortunes exceed one billion dollars, with a new billionaire every two days. (Oxfam 2018)[6]

A critical statement by Niose (2011) defines the "phenotype" of corporate business goals:

> Corporations are the only entities that actually have millions of dollars at their disposal for the sole purpose of defending their economic interests, shaping public policy, utilizing the courts, controlling the media, and advertising themselves and their products. With vast economic resources, no moral sense, and the singular goal of making money, corporations act narcissistically as a matter of ordinary practice. Even when they act charitably, they do so as a matter of public relations and some kind of business interest. The libertarian guru Milton Friedman, leaving all niceties aside, has expressly condemned the notion of corporate charity as a breach of duty to shareholders, unless it can be clearly justified for reasons of public relations or the like. (Niose 2011)[7,8]

Competitive market conditions and greed are culturally fueling corporative and national interests' behaviors into a risky future. This also impacts the evolution of the Earth's biocapacity, according to the Footprint Network (2019) (Fig. 11.2). How corporate greed is fueling these projects cannot be better summarized than by the following statements:[9]

> Peter Diamandis paraphrased Lewis at the 2006 International Space Development Conference when he exclaimed, There are twenty-trillion-dollar checks up there, waiting to be cashed!
>
> This $20 trillion figure is based on Lewis's calculations of how much a metallic asteroid (3554 Amun) would be worth if it was sold at current market prices.

Additionally, the column published in the journal *Nature* (World View) completes the current state of mind in this issue:

> Let's mine asteroids — for science and profit.
>
> Asteroid mining is a capitalist exercise, but the financial and technical barriers to entry are high. Just as the US government encouraged the development of the American West in

[5] https://www.credit-suisse.com/corporate/en/responsibility/news-stories/articles/news-and-expertise/2015/10/en/global-wealth-in-2015-underlying-trends-remain-positive.html

[6] https://www.oxfam.org/es/sala-de-prensa/notas-de-prensa/2018-01-22/el-1-mas-rico-de-la-poblacion-mundial-acaparo-el-82-de-la. https://www.bbc.com/mundo/noticias-42776299

[7] https://thehumanist.com/magazine/may-june-2011/features/corporate-power-and-todays-humanist

[8] https://www.nature.com/news/space-miners-seek-riches-in-nearby-asteroids-1.10513

[9] http://space.nss.org/book-review-mining-the-sky/

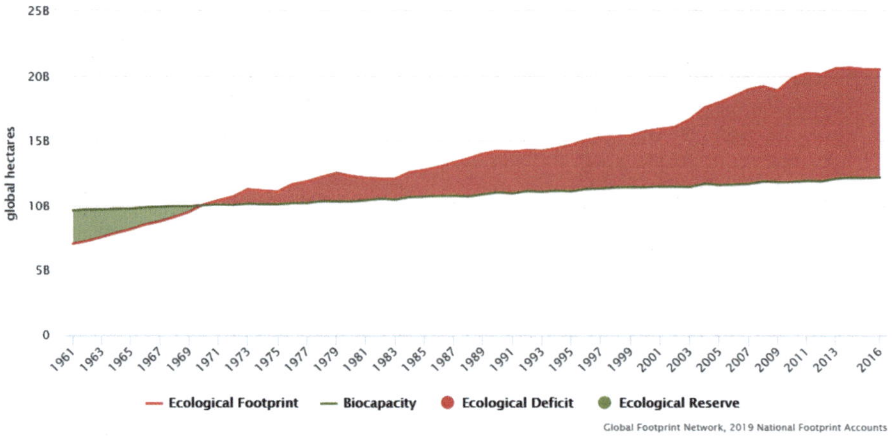

Fig. 11.2 Timeline of the relationship between biocapacity and ecologic footprint, resulting in an increasing ecological deficit. "*Ecological footprint vs biocapacity measured in global hectares (Global Footprint Network, 2019)*". https://www.researchgate.net/publication/337907452_Product-Service_Development_for_Circular_Economy_and_Sustainability_Course_Training_Modules_Introduction_to_the_circular_economy/figures?lo=1

> the nineteenth century, so NASA and space scientists can help to tame the frontier of space, and make it ripe for harnessing, to the mutual benefit of all involved.
>
> To promote such research, NASA's new goal should not be exploration, but enablement of the commercial development of space resources. Exploration will follow naturally. (Elvis 2012)

Therefore, one fundamental problem relates to the profit of the industrial and financial corporations and geopolitical interests, whether at the Earth ground level or extraterrestrial. Under this greedy profit perspective, it seems impossible to separate profit and greed from developmental strategies and shared interests when corporative goals establish these. Breaking such present-day close association seems to call for a significant cultural departure, as mentioned in "Humanism in Business–Towards a Paradigm Shift?" (Prison and Lawrence 2010). This objective seems close to a utopian goal since the world cultural condition implies multiple variants that represent the cultural richness of our species. It would be counterfactual to pretend imposing universal progress criteria and cultural living welfare conditions in this crackled, uneven, reality. However, some goals will be pursued stemming from technologically advanced communities, and the already set up competitive power race among a group of nations and corporative profit-seeking interests.

References

Barrett JL. Homo Symbolicus, edited by C.S. Henshilwood and F. d'Errico, Amsterdam/Philadelphia: John Benjamins Publishing Company, Chapter 11, 2011.

Botero CA, Gardner B, Kirby KR, Bulbulia J, Gavin MC, Gray RD. The ecology of religious beliefs. PNAS. 2014;111:16784–9.

Boyer P, Lienard P. Behav Brain Sci. 2006;29:1–56.

Colombo JA. Pobreza y desarrollo infantil (Poverty and child development). Ed. Paidos, Buenos Aires, 2007.

Colombo JA. Our Animal condition and social construction. New York: Nova Science Publishers; 2019. ISBN: 978-1-53615-357-6. eBook ISBN: 978-1-53615-3583

Colombo JA. The Homo within the sapiens. New York: Nova Science Publishers; 2021.

Elvis C. Let's mine asteroids for science and profit. Nature. 2012;485:549. https://doi.org/10.1038/485549[a].

Harari YN. *De Animales a Dioses*. Bogotá: Penguin Random House; 2015.

Henrich J, Boyd R, Derex M, Kline MA, Mesoudi A, Muthukrishna M, Powell AT, et al. Understanding cumulative cultural evolution. PNAS 113: E6724-E6725, 2016. https://www.pnas.org/content/113/44/E6724.

Inzlicht M, Tullet AM, Good M. Relig Brain Behav. 2012;1:192–212. https://doi.org/10.1080/2153599X.2011.647849.

Mithen S. The prehistory of mind: a search for the origin of art, religion, and science. London: Thames and Hudson Ltd.; 1996.

Montardy Q, Kwan WC, Mundinano IC, Fox DM, Wang L, Gross CT, Bourne JA. Mapping the neural circuitry of predator fear in the nonhuman primate. Brain Struct Funct. 2021;226:195–205. https://doi.org/10.1007/s00429-020-02176-6.

Peoples HC, Marlowe FW. Subsistence and the evolution of religion. Hum Nat. 2012;23:253–69. https://doi.org/10.1007/s12110-0129148-6.

Powell A, Shennan S, Thomas M. Late Pleistocene demography, and appearance of modern human behavior. Science. 2009;324:1298–301. https://doi.org/10.1126/science.1170165.

Prison MA, Lawrence PR. Humanism in business – towards a paradigm shift? J Bus Ethics. 2010;93:553–65. https://doi.org/10.1007/s10551009-0239-1.

Sidanius J, Pratto F, van Laar C. Social dominance theory: its agenda and method. Polit Psychol. *2004*;25:845–80.

Vaesen K, Collard M, Cosgrove R, Roebroeks W. Population size does not explain past changes in cultural complexity. Proc Natl Acad Sci USA. 2016;113: E2241-E2247.

Wilent WB, Oh MY, Buetefisch CM, Bailes JE, Cantella D, Angle C, Whiting DC. J Neurosurg. 2010;112(112):1295–8. https://doi.org/10.3171/2009.9.JNS09577.

Wiltermuth SS, Heath C. Assoc Psychol Sci. 2009;20:1–5.

World Bank Group. Poverty and Shared Prosperity 2020. Reversals of Fortune, page 39.

Chapter 12
On the Human Dimension

On Artificial Intelligence (AI); Cyborgs: To Be or Not to Be?...

> .... we should ask ourselves whether we should create sentient AI beings when we can't even fulfil ethical obligations to the sentient beings already on the planet. If AI is to best support human, do we want to create beings that we have ethical obligations to, or mindless AIs that make our lives easier?[1]

The issue of consciousness invades several aspects of natural and artificial existence (AI, cyborgs, robotics) (Fig. 12.1). A glimpse of the debate around this issue is given in several articles by Hildt (2019) and Schneider (2019).[2] Avoiding entering the debate regarding the existence of consciousness in the animal kingdom and the human creative artificial beings, it seems adequate to state a few premises that may help focus this complex issue, permeable to anthropocentric interpretations.

Thus, if we consider self-awareness a component of any definition or concept construct of consciousness, in terms of relative biological complexity, a hypothetical behavioral scale would start from reflex responses induced by environmental changes as it takes place in unicellular organisms. This "reactive awareness" of attractive or repulsive responses to environmental changes (physical or chemical) is a universal characteristic of cell structures, whether in uni- or pluricellular entities. In more complex organisms, environmental and internal body conditions (hunger, fear, cold, pain, arousal) construct an interactive integer of self-comfort or

[1] Library of Congress, J. W. Kluge Center. https://blogs.loc.gov/kluge/2019/10/will-ai-become-conscious-a-conversation-with-susan-schneider/. Interestingly, Schneider voices a concerning warning: "Several large research projects are currently trying to put AI inside the brain and peripheral nervous system. They aim to hook you to the cloud without the intermediary of a keyboard. For corporations doing this, such as Neuralink, Facebook and Kernel, your brain and body is an arena for future profit. Without proper legislative guardrails, your thoughts and biometric data could be sold to the highest bidder, and authoritarian dictatorships will have the ultimate mind control device. So, privacy safeguards are essential."

J. A. Colombo, *Dominance Behavior*, https://doi.org/10.1007/978-3-030-97401-5_12

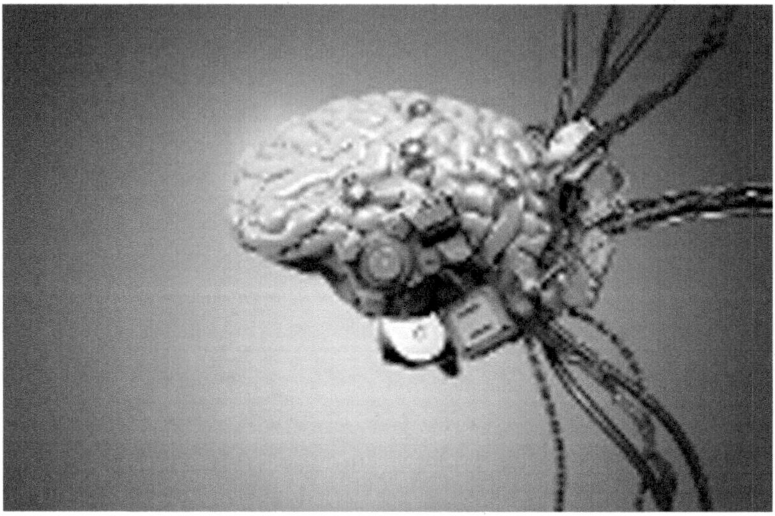

Fig. 12.1 High-tech brains...? (Phonlamai Photo Getty Images)

discomfort, pleasure, or demand. This induces the corresponding behavior (seeking protection, attack, flight) to satisfy them. Depending on animal orders, for example, flight and attack imply the construct and execution of a given strategy to secure survival or dominance, which interacts with experience and, thus, is adaptive.

These elements of behavior imply sensing physical, chemical, or emotional components. Hence, sensory (interoceptive and exteroceptive) and perceptive inputs compose the basic grounds on which progressive neural net complexity conditions the behavioral menu.

This opens a debatable issue on whether behavior implies a conditional trigger interactive with a recorded and stored historical precondition, that is, a memory of previous experiences, whether of pleasure or discomfort or life-risk character. In this context, "initiative" or intentional behaviors could be consequences of a complex interactive net that involves personal history and environmental triggers of behavior; this would suggest the existence of a proto-conscience or self-conscience-like in living species beyond the involved condition of reflex stimulus reaction.

Regarding AI and human contexts, several issues emerge, including:

> An insufficient grasp of the nature of self, consciousness, and mind could undermine the use of AI and brain enhancement technology, bringing about the demise or suffering of conscious beings. To flourish, we must grasp the philosophical issues lying beneath the algorithms.[2]

> Several large research projects are currently trying to put AI inside the brain and peripheral nervous system. They aim to hook you to the cloud without the intermediary of a keyboard. For corporations doing this, such as Neuralink, Facebook and Kernel, your brain and body is an arena for future profit. Without proper legislative guardrails, your thoughts and

[2] https://www.scientificamerican.com/article/who-wants-to-be-a-cyborg

> biometric data could be sold to the highest bidder, and authoritarian dictatorships will have the ultimate mind control device. So, privacy safeguards are essential. I have other worries as well. I worry that people will feel pressured to use brain implants so that they can stay competitive in their jobs or college... In addition, I'm concerned that if only the elite few are enhanced there will be a vast intellectual gulf between the have and have-nots. (Schneider 2020)[3]

The challenges that bring artificial intelligence developments are perhaps better described in Koch's proposal (2017):[4]

> To keep up with AI, we'll need high-tech brains. (Fig. 12.1)

On Environmental Degradation

Profit-seeking and financial corporative empowering and demands for the basic needs of impoverished populations converge to construct a degrading horizon of our ecological environment, including citizen awareness and environmental education. They include climate change, land degradation, air and water pollution, and species extinction, examples of the widespread impact level of human Anthropocene. Its consequences are linked to rampant consumerist behavior, profit-seeking, and media pressure on collective citizen behavior that downplay—for any effective practical purposes—consideration on public education aimed at environmental care.

Humans impact the physical environment in many ways: overpopulation, pollution, burning fossil fuels, and deforestation. Changes like these have triggered climate change, soil erosion, poor air quality, and undrinkable water. These negative impacts can affect human survival and behavior and prompt mass migrations or battles over clean water and living territory.[5] As stated by Wackernegal et al. (2021):

> **As humanity's demand on natural resources is increasingly exceeding Earth's biological rate of regeneration, environmental deterioration such as greenhouse gas accumulation in the atmosphere, ocean acidification, and groundwater depletion are accelerating. As a result, the capacity of ecosystems to renew biomass, herein referred to as 'biocapacity', is becoming the material bottleneck for the human economy. Yet, economic development theory and practice continue to underplay the importance of natural resources, most notably biological ones.** (Bold type inserted by J.A.C)

---

[3] https://www.newscientist.com/science-events/artificial-intelligence-susan-schneider/?utm_source=nscon&utm_medium=email&utm_campaign=02052021_big_thinker_ai&

[4] C. Koch. https://www.wsj.com/articles/to-keep-up-with-ai-well-need-high-tech-brains-1509120930

[5] https://www.nationalgeographic.com/magazine/article/helmeted-hornbill-bird-ivory-illegal-wildlife-trade?cmpid=org=ngp::mc=crmemail:src=ngp::cmp=editorial::add=Compass_20210501&rid=95527DDF86250D8AB073474627963407

In addition to previously published critique on management of world matters (Colombo 2007, 2010, 2019, 2020), the following brief mentions on contamination effects on species living on remotely different ecosystems underscore the dimension of the impact of human consumerism and disregard for the health of our shared planet Earth.

In describing mining and hydrocarbon exploitation in the arctic territory, Pilar Bonet (2019)[6] provides a further example of how our planet is being managed by wealthy individuals and corporations, which reinforces the concept of a false sense of globalization regarding decisions on our ecosystem. This description suggests some form of corporative or financial autocracy. The report from UNICEF (2020)[7] describes the other side of the world condition:

> Almost 690 million people around the world went hungry in 2019. As progress in fighting hunger stalls, the COVID-19 pandemic is intensifying the vulnerabilities and inadequacies of global food systems. While it is too early to assess the full impact of the lockdowns and other containment measures, at least another 83 million people, and possibly as many as 132 million, may go hungry in 2020. The setback throws into further doubt the achievement of the Sustainable Development Goal for zero hunger.

This adds to the following reports (W. Johnson note on P. Salopek report, "When drinking water disappears," National Geographic, 2020):[8]

> Some 600 million people in India already live without clean water. Industrial waste, urban sewage, and agricultural runoff has poisoned entire river systems. The Ganges, venerated by Hindus, is one of the most polluted rivers on Earth.
>
> Contamination of fish by mercury, PCBs and other toxic chemicals appears to be increasing in the USA (NRDC, 1998) and evidence about the effects of low- level but possibly wide-spread contamination of fish is increasing in Europe (Matthiessen, 1998; Tyler, 1998). Recent results from the UK, for example, suggest that the incidence of feminization and other sexual disruption in fish 'is higher than previously thought and is associated with discharges from sewage treatment works' (EA, 1998) (European Environmental Agency, 2021).[9]

Due to the scientific and technological development of our species, a progressive gap developed in two domains: first, the financial resources poured into extraterrestrial exploration and colonization for mining purposes and corporative profit, from a planet ecologically placed at high risk and with degraded living conditions, and second, the distorted national realities in terms of institutional rights in the face of:

[6] https://elpais.com/elpais/2019/05/21/eps/1558434406_575807.html

[7] https://data.unicef.org/resources/sofi-2020/?utm_source=newsletter&utm_medium=email&utm_campaign=nutrition_SOFI_2020

[8] https://www.nationalgeographic.com/magazine/2020/08/indias-daunting-challenge-there-is-water-everywhere-and-nowhere-feature/?cmpid=org=ngp::mc=crm-email::src=ngp::cmp=editorial::add=Photography_20200717&rid=95527DDF86250D8AB073474627963407

[9] https://www.eea.europa.eu/publications/NYM2/page008.html

> 72% of the world population live in countries faced with a precarious situation. These countries both (1) run a biological resource deficit (where demand for biological resources exceeds regeneration) *and* (2) generate less than world-average income, limiting their ability to purchase resources from elsewhere. (Wackernagel et al. 2021)

This is further considered in the following paragraphs.

Uncontrolled human activities result in various domains of debris production and contamination. They involve conditions of poverty and marginalization that sequester vast populations that remain outside of human development horizons and acceptable living conditions. Additionally, physical and chemical debris contaminate air, ground, and water domains, as well as extraterrestrial space (Fig. 12.2). Is free-running profit the fundamental motivation that will mold our human society into a path toward equal rights and place cognitive enhancement, access to information, and well-being within the reach of vast, impoverished, populations?

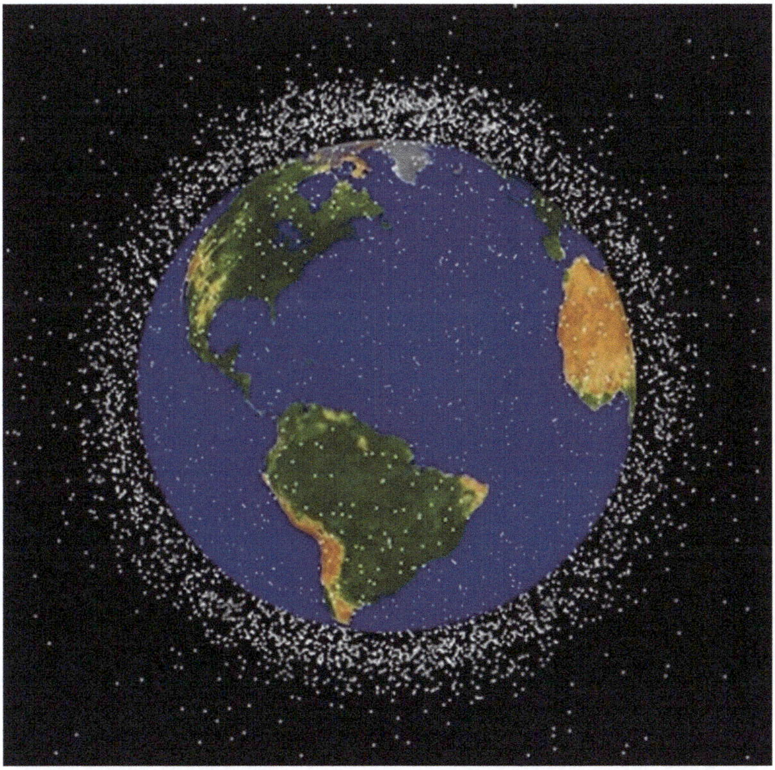

Fig. 12.2 This NASA graphic depicts the amount of space junk currently orbiting Earth. The image depicts objects at low Earth orbit and is the region of space within 2000 km of the Earth's surface. It is the most concentrated area for orbital debris. (Credit: NASA ODPO; https://orbitaldebris.jsc.nasa.gov/photogallery/)

In summary, our species has entered a high-risk path that menace our survival and Earth ecological balance, in terms of technological competition for dominance/prevalence. Where is the limit...?

Inequality

> 'Inequality' –or 'natural-inequality' expressed in gregarious animal communities with a hierarchical structure– raises an issue of 'fundamental animal conditionality' as 'naturally inserted' in human social structures. The central question remains on how to bypass such ancestral trend, and to evaluate its impact on the members of the different communities, and on global interaction. (Colombo 2019)

Throughout natural history, animal behavior was based on resources that could be taken—regarding feeding or developmental and reproductive advantages—and defended. Projected onto *Homo sapiens* behavior, it boils down into universal cultural domains: dominance/prevalence (eventually expressed as authoritarianism; caste; socially and physically degraded populations due to chronic, harsh inequalities; or class privileges) and profit (eventually transformed into greed and developmental opportunity inequalities), including all forms of human slavery as depicted in the following statements:

> …new estimates on child malnutrition indicate that 149 million children under 5 were affected by stunting, 45 million suffered from wasting and 39 million were overweight in 2020. And these estimates do not yet account for the impact of the pandemic.[10]

> Inequality within world regions varies greatly. In 2016, the share of total national income accounted for by just that nation's top 10% earners (top 10% income share) were 37% in Europe, 41% in China, 46% in Russia, 47% in US-Canada, and around 55% in sub-Saharan Africa, Brazil, and India. In the Middle East, the world's most unequal region according to our estimates, the top 10% capture 61% of national income.[11]

> In recent decades, income inequality has increased in nearly all countries, but at different speeds…[12]

> It's the first time five people have had fortunes in their USD 100 billion respectively, according to Bloomberg's Billionaires Index, a daily wealth ranking. French luxury tycoon

[10] https://data.unicef.org/resources/jme-report-2021/

[11] World Inequality Report, https://wid.world/document/world-inequality-report-2018

[12] https://www.infobae.com/america/agencias/2020/11/10/euforia-bursatil-crea-record-de-fortunas-de-us100000-millones

> Bernard Arnault joined Bezos, Bill Gates, Mark Zuckerberg, and Elon Musk to be part of this exclusive club.[13]

Besides the rampant social inequalities, concentrated fortunes derive on political power distribution and impact on the development decision coordinates of our world community.

Thus, sustained nutritional and cognitive inequalities revert into diverse curves of individual maturation and access to social, physical, and emotional quality of life, as characterized by comparison with optimal developmental processes and productive/creative social insertion. Cognitive comparative developmental handicap resulting from profoundly unequal socioeconomic conditions marks the probability of constructing either optimal or degraded cognitive development and performance. Social insertion, social mobility, social structure, and the spectrum of accessible social (labor) roles are deeply affected by early inequalities during development. As mentioned in previous chapters, these profound inequalities project socially as coercive developmental conditions that mimic those observed in other animal species.

Our species should evolve along humane routes, i.e., providing equal access to physical and cognitive development and promoting solidarity and merit distributed along with the various social domains. Hence, profit should prioritize cultural values, implying a shift in our cultural structure. In this context, merit should not imply a financial social privilege or political power as a necessary consequence. Other means should provide recognition to special social contributions. Profit predominance in our social values and social behavior leads to a spin-off of inequalities, power concentration, and political dependence dislodged from collective enhancement and constructive public cognitive alert.

Not alien to this social crack, individual space flight[12, 13] is announcing a new era but at the same time stressing the deep gap in our human society, amid social and cognitive profound inequalities and the need for new strategies and financial efforts:

> Space isn't just for the professionals anymore, as several high-profile rocket-makers are gearing up to send civilians above the atmosphere. But with price tags in the millions, we are still far from the long-awaited democratization of space flight. Many of these civilian space flight opportunities are being run as contests, auctions or raffles. *Blue Origin* is auctioning off a seat aboard its very first crewed flight on the *New Shepard* suborbital rocket – at time of writing, the price had reached $2.8 million.[14]

The state of the world in terms of several domains, as described, suggests the question raised earlier (Colombo 2007), "Are we the wrong species?," and the question that follows: can we succeed in breeding a subspecies with different values, somewhat diverted from ancient animal drives based on dominance...?

---

[13] https://www.newscientist.com/article/2278309-more-people-are-going-to-space-but-diversity-is-still-grounded/#ixzz71HcC5Dg5

[14] Ibid.

Ancestral Hierarchy and Conflict

The unequal development of nations, war power, financial support dependence, technological development, sociopolitical structures and religious beliefs, and collective genotypical behaviors assure that equilibrium represents a distant goal to achieve. Additionally, the animal kingdom history provides grounds to speculate that thriving for supremacy to secure nutrients, a position of dominance/prevalence, and survival is embedded in the core of all natural beings, including humans. To these, *Homo sapiens* has superimposed layer after layer of cultural dialectics that give grounds to construct the illusory belief that those previously mentioned drives no longer apply to us, drowned in human culture.

However, human history and humankind behavior on a day-to-day basis provides arguments to the contrary. We are built with ancient drives that originated in the unicellular basic responses of thrive or flight, survive or perish, now masked under several layers of complex, often hypocritical cultural behaviors. It appears that the only way out is the progressive, constant merge of human "subspecies" specimens supporting radically different behaviors that think and act under new sets of values that could balance our ancient animal history and progressively reduce social inequalities. Perhaps hidden signs of such battle have been historically underway in different forms involving individual actions and freedom movements. Yet, until (and if) that time comes, it will be the battle of all battles where present generations would be absent. The quest for survival will predominate in our social behaviors in the form of massive responses to inequity in living conditions and access to cognitive development. In the meantime, technological and financial elites will find their getaway on Earth or elsewhere.

> The massacre site from Potočani (*6,200-year-old;* east Croatia) is not the first found from European prehistory… The only thing that is abundantly clear is that this fundamentally dark human behavior has persisted for millennia. Mass killings have taken place all over the world for at least 13,000 years.[15]

This statement announces just what chimpanzees and other species also have:

> Today humans retain the genetic underpinnings of both conflict and conflict management; thus, we retain the potential for both war and peace. (Boehm, 2012b)

As stated before (Colombo 2019), creativity, nonkin cooperation, and solidarity will provide conditions for equal opportunity cognitive development. That should supply us with the fundamental characteristic profile of our future human species.

Harsh living inequalities hamper collective human development and undresses a dark component of an often-voiced ultra-sociality character of our species, as optimistically stated by Buckholtz and Marois (2012):

---

[15] https://www.nationalgeographic.com/history/article/dna-study-ancient-massacre-victims-raises-more-questions-answers?cmpid=org=ngp::mc=crm-email:src=ngp::cmp=editorial::add=Compass_20210619&rid=95527DDF86250D8AB073474627963407

> … *Homo sapiens* is unique in its capacity for widespread cooperation and prosocial behavior among large and genetically heterogeneous groups of individuals. This ultra-sociality figures largely in our success as a species…

This statement neglects the historically proven human capacity for cruelty (Colombo 2021), parochiality, physical and economic oppression, and environmental neglect, a worrying set of behavioral profiles expressed throughout human history and practiced by leaders and followers of the most diverse political, religious, and financial currents. Shall human species end as an institutionalized evolution based on technological dominance, or based on castes, or profoundly stratified social inequalities generated by unequal access to cognitive development? Or, alternatively, as a solidarity ecologically balanced community that provides new meaning to values such as merit, universal basic well-being,[16] solidarity, and political and financial priorities based on actual collective progress? Or, as Chomsky (2021) states in *Cooperation or Extinction*, shall our species persist in universal threats that represent non-curved climate change and nuclear developments by warmongers placing humanity at risk, a concept shared by the Anthropocene Working Group and the Bulletin of the Atomic Scientists?

The Issue of Gene-Culture Coevolution and Behavioral Spin-Offs

In discussion with Vaesen et al. (2016), Henrich et al. (2016) states that a significant relationship between census population size and cultural complexity depends on the size of the population that shares information. That is, the adequate cultural population size is what matters. Ideologists have taken this concept to support limited, biased ideas of political representation of what *true* representative democracy should be, which deals with several issues. But this complex issue merits an in-depth discussion beyond the objective of this essay.

Gene-culture interaction has set up an unsolved discussion about whether culture can modify the genetic structure or modulate its social expression and adaptive process. In the case of critical conflicts between both behavioral domains, social pathologies ensue. Hence, it is underlined that individual or collective pathologies tend to arise whenever social structures violate natural needs.

Richerson (2017) exemplifies the adaptive cultural role—among other examples—with Inuit metabolic adaptation to feeding resources and states that culture sets up the environment where genes evolve and adapt:

---

[16] "Five years into the race towards the 2030 Sustainable Development Goals, the world is not on track to achieve universal access to safely managed drinking water and sanitation and basic hygiene services, according to the latest WHO/UNICEF JMP report, Progress on household drinking water, sanitation and hygiene 2000–2020: Five years into the SDGs." https://data.unicef.org/resources/progress-on-household-drinking-water-sanitation-and-hygiene-2000-2020/?utm_source=newsletter&utm_medium=email&utm_campaign=JMP%202021

> Cultures substantially construct the environments in which genes evolve.

However, the opposite could be operative, i.e., cultures evolve according to genetic and ecological interactions, and individual/collective imprints. Thus, the mentioned statement conflicts with the concept that behavioral expressions (including cultural) result from genetic and socioecological interactions, including individual and collective history of phenotypic adaptations conditioned by ancient drives (see also Neuberg et al. 2010). Culture is not the only factor in setting up environmental changes where individuals adapt nor in establishing individual physiological (brain and mental) and somatic characteristics and adaptive endurance threshold. Though social enclosures modulate genetic expression, the question arises when socioeconomic conditions invade the domain of basic survival needs imprinted in genetic construction; in this case, conflict arises. At this point, either reactive responses ensue or social pathology does, indicating that socialization—institutionalization—does not erase basic, ancestral survival drives. It may transiently adapt, block their expression under repressive circumstances, or enter conflictive tensions. It also takes place under conditions where essential feeding resources are compromised or when political oppression empties, menace, or cancels physical and emotional response potential. This is observed in isolated, marginal, underfed—when chronic, a condition that compromise physical and social vigor—groups or, as history shows, under repressive political circumstances. It also develops under religious, mystical, or altered consciousness conditions. Sublimation—in general, diversion—of basic drives has been invoked by psychological theories to represent a product of coercive institutionalization process. In strongly stratified or repressive societies, the latter process—diversion of basic needs and drives—does not operate with identical social reactivity thresholds and timing nor induce identical behavioral expressions compared to traditional democratic societies.

The Institutional and Cultural Carcass of Ancient Animal Drives

> Humans are animals. As such, human brains, like the brains of all animals, evolved via natural selection to solve the types of recurring fitness-relevant problems that our ancestors faced long ago. (Neuberg et al. 2010).

> Culture is one of the two or three most complicated words in the English language. This is so partly because of its intricate historical development, in several European languages, but mainly because it has now come to be used for important concepts in several distinct intellectual disciplines and in several distinct and incompatible systems of thought. It is especially interesting that in archaeology and in cultural anthropology the reference to culture or a culture is primarily to material production, while in history and cultural studies the reference is primarily to signifying or symbolic systems. This often confuses but even more often conceals the central question of the relations between 'material' and 'symbolic' production… (Williams 1976)

> It is clear that throughout the history of anthropology, scholars have adapted their notions of culture to suit the dominant concerns of the day, and they will no doubt continue to do so. Little is to be gained, therefore, from attempts to legislate on the proper meaning of the term. (Engold 2002)

If we consider culture as a system of knowledge and social values shared by groups of people, with their material objects and possessions, under an evolutive perspective, it developed mounted on subdued ancient behavioral drives that stem from our animal origin. From a neuroscientific perspective, the latter exists in basal brain circuits that subserve behaviors involving fear, physical and social survival needs, and reproduction. These drives provide the basement to more complex and evolving constructions derived from experience, inventive, and social sharing, involved in our concept of culture.

The evolution of symbolic language, virtual constructions, creativeness, technological development, and Earth-bound species dominance provided *Homo sapiens* of arguments to construct the illusion of our departure from animal heritage. Yet, ancient animal drives deeply anchored in our original neurobiological construct are revealed in our search for dominance (whether on social, political, or warfare grounds) and secure survival. These behaviors are expressed as an individual or collective profile that eventually submit third parties to oppression or torture, unseen in the natural kingdom. They are also present in other social behaviors, such as crime rates, mass murders, and institutionalized privileges that crystallize social inequalities (access to basic living conditions, access to cognitive development). Additionally, it is expressed in the sophisticated use of media and material conditions to bend collective resistance and, hence, promote submissive behaviors and cancel creative minds, sometimes applied to economically, nutritionally, and cognitively deprived populations.

> In 2020, one in four people in the world (2 billion people) still lacked safely managed drinking water services, nearly half the world's population (3.6 billion people) lacked safely managed sanitation services, and 2.3 billion people did not have access to basic hygiene services.[17]

Could our species progressively dismantle or replace –or free our behavior– from ancient, primeval animal drives that ground us to parochial and hierarchical behaviors or long-term adaptations to past Homo conditions? Neural circuit reconfiguration following sociocultural experiences (e.g., by Issa et al. 2012) involves recent neural processing rather than basic, ancient, survival-related behavioral circuits. Under critical survival conditions ancient processing modes seem to override more recently acquired ones.

The following questions, thus, would seem to require profound neurobiological and mental reconfiguration. Could cooperative behavior and a culture of solidarity unplug our dependence from such primeval drives? Could we reconfigure

---

[17] https://www.who.int/news/item/01-07-2021-billions-of-people-will-lack-access-to-safe-water-sanitation-and-hygiene-in-2030-unless-progress-quadruples-warn-who-unicef

fundamental ancient drives through education aimed towards rational decisions that prioritize collective rights and developmental values? Could we reconfigure basic, ancestral neural circuits, or is social plasticity operating only on circumstantial or short-term behaviors? Otherwise, how could we explain our species warmonger and brutal behaviors.

According to some authors, perhaps the secret of our species' success does not lie on our collective profile but on its social, cognitive, and financial stratification, which would represent an oxymoron of our best sense of equal rights.

Finally, should we rethink—reformulate—what being human should mean? An answer would require profound changes in cultural, political, and financial power structures in order to prevail over our ancient animal drives that handcuff our species within the belt of the universals of animal behavior.

In conclusion:

> Highly stratified information access, the arsenal of war, the demands of a runaway economy–with periodic financial crises– and the inability to self-sustain without resorting to policies of 'adjustment' and overexploitation of natural and human resources, while lacking efficient and sustainable ways to dispose of our disposals, constitute a set of strategies that do not build a promising future for world majorities, nor for communities attached to ethnic traditions, nor for the planet. (Colombo 2019)

Worldwide, our species' complex cultural composition implies a culturally fractured condition, variety of phenotypes, cultural profiles, and social demands. As western societies evolved from intricate, interactive hierarchical actions and structures (privileges, conquest, fanaticism, racism, civil wars, religious persecution), others would do from their original, historical cultural patterns. So, what opportunities are there to pursue a sustainable cultural variety? Within the dynamic history of social values, the unyielding exception implies protection of human lives on both physical and intellectual domains, cultural profiles, well-being, access to knowledge, and progressive, dynamic, widespread evolution toward these basic domains.

Fueled only by profit goals and social prevalence tends to contradict the above goals and dampen our species capacities to evolve as a multicultural species and to avoid generating conditions for poverty, marginality, and ecosystem destruction. Unless our species reaches a sustainable, viable multicultural condition, we will have to conclude that we became the wrong surviving *Homo* species despite technological and cultural developments (Colombo 2010).

References

Buckholtz JW, Marois R. The roots of modern justice: cognitive and neural foundations of social norms and their enforcement. Nat Neurosci. 2012;15:655–61.

Chomsky N. Cooperación o Extinción (Cooperation or Extinction) (Original title in First edition, Internationalism or Extinction). London: Ed. Penguin Random House; 2021.

Colombo JA. Pobreza y desarrollo infantil (Poverty and child development). Buenos Aires: Ed. Paidos; 2007.

Colombo JA. Somos la especie equivocada? (Are we the wrong species?). Buenos Aires: Ed. EUDEBA; 2010.

Colombo JA. Our animal condition and social construction. New York: Nova Science Publishers; 2019. ISBN: 978-1-53615-357-6. eBook ISBN: 978-1-53615-3583

Colombo JA. Creativity, a profile for or species. Newcastle: Cambridge Scholars Publishers; 2020.

Colombo JA. The Homo within the sapiens. New York: Nova Science Publishers; 2021.

Engold T. Introduction to culture. In: Company on encyclopedia of anthropology. 2nd ed. London: Routledge; 2002. https://www.taylorfrancis.com/chapters/edit/10.4324/9780203036327-19/introduction-culture?context=ubx.

Henrich J, Boyd R, Derex M, Kline MA, Mesoudi A, Muthukrishna M, Powell AT, et al. Understanding cumulative cultural evolution. PNAS. 2016;113:E6724–5. https://www.pnas.org/content/113/44/E6724

Hildt E. Artificial intelligence: does consciousness matter? Front Psychol. 2019;10:1535. https://doi.org/10.3389/fpsyg.2019.01535.

Issa FA, Drummond J, Cattaert D, Edwards DH. Neural circuit reconfiguration by social status. J Neurosci. 2012;32:5638–45.

Neuberg SL, Kenrick DT, Schaller M. Evolutionary social psychology. In: Fiske ST, Gilbert DT, Lindzey G, editors. Handbook of social psychology. Wiley; 2010. p. 761–96.

Richerson PJ. Cultural evolution and gene-cultural coevolution. Evol Stud Imag Cult. 2017;1:89–92.

Vaesen K, Collard M, Cosgrove R, Roebroeks W. Population size does not explain past changes in cultural complexity. Proc Natl Acad Sci U S A. 2016;113:E2241–7.

Wackernagel M, Hanscom L, Jayasinghe P, Lin D, Murthy A, Neill E, Raven P. The importance of resource security for poverty eradication. Nat Sustain. 2021; https://doi.org/10.1038/s41893-021-00708-4.

Williams R. Keywords: a vocabulary of culture and society. New York: Oxford University Press; 1976.

Chapter 13
Our Species' Hypothetical Alternatives

Our Species' Hypothetical Alternatives. Need for a Basic Cultural Change

Within the *universals of behavior*, previous chapters dealt with dominance and survival behaviors as ancient neurobehavioral animal components with their processing neural circuits predominantly placed in basal brain structures. Phylogenetically newer brain neural circuits and structures evolved on top of them, performing higher processing of behavioral domains, social values and interactions, and subjected to ultimate survival strategies, as stated by Dawkins in *The Selfish Gene* (1995).

The threat imposed by the human species demands on the sustainability of our ecological environment and biosphere capacity has been clearly stated by Kitzes et al. (2008):

> Despite ample recognition of the importance of achieving sustainable development, exemplified by the Rio Declaration of 1992 and the United Nations Millennium Development Goals, the global economy fails to meet the most fundamental minimum condition for sustainability –that human demand for ecosystem goods and services remain within the biosphere's total capacity.

Thus, ecological conservation represents one main condition for our species survival (see Fig. 13.1 for world distribution of biocapacity). It depends on the cultural profiles of the different human settlements and developments around the world. More specifically, they depend on the strategies and goals of prevailing corporative strategies toward profit and generation of human living standard needs. Political and corporative interests insert and promote a consumerist culture in the collective social body. This leads to increasing demands on nonrenewable extraction of mineral resources and water reserves, water and air contamination, and competitive strategies to prevail on the profit from such precious resources.

A hypothetical survival equation that could satisfy the multiple cultural expressions and vested financial interests would involve several variables of a problematic

J. A. Colombo, *Dominance Behavior*, https://doi.org/10.1007/978-3-030-97401-5_13

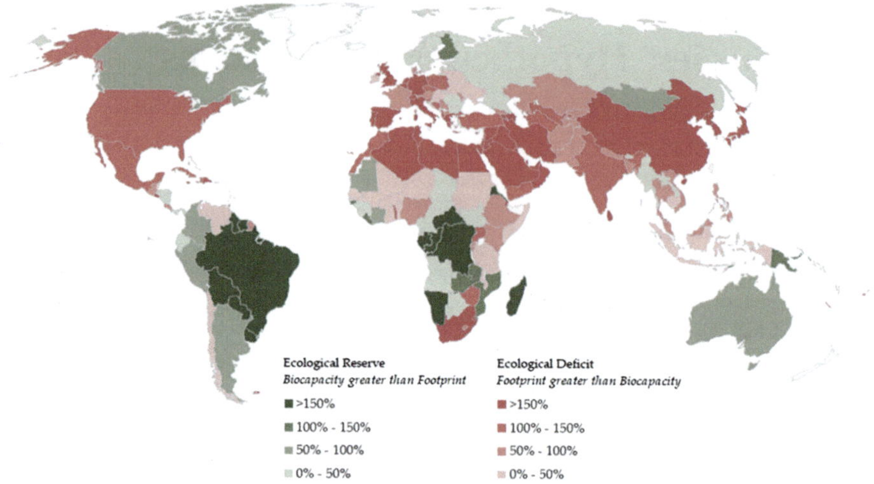

Fig. 13.1 Map of countries with biocapacity deficit (pink and reddish tones) and reserves (green tones) in year 2014, according to the Global Footprint Network 2018. (*Color modified*) See also in Lin et al. (2018). (Source: Global Footprint Network, www.footprintnetwork.org)

solution. Consumerist and profoundly unequal living, health and education conditions, and social marginality should be modified. These depend on skewed access to technological development, unleashed material profit, corporative grasp of collective needs, needed sociopolitical development, damped cognitive alertness, and educational inequalities. Given the variety of parochial cultures and wealthy interests, to start with, it would be necessary to find common, basic, value denominators, probably linked with equal opportunity for cognitive development and labor opportunities within any given cultural construct.

For world financial elites, it could be applied the metaphor of bee colonies searching for other settlements. This, in modern times, aimed toward extraterrestrial horizons after exhausting—compromising—terrestrial resources. Reduced job opportunities and feeding resources and overpopulation promote human migration movements, social detachment, and new demands. In these socially tense conditions, it seems proper to recall Frans de Waal (2005) statement that the members of our species would be behaviorally characterized as bipolar characters. Furthermore, he recalls an old Roman proverb:

> ["*Homo homini lupus*" ("man is wolf to man"), captures this asocial vision that continues to inspire law, economics, and political science today.]

The above statement poses a further question, considering the intra- and intercommunity relationships with massive migratory movements, for the history of humankind is overly demonstrative of rejection and prejudice behavior—if not manifest dominance and oppression affecting several individual developmental domains—for human cooperation and solidarity are conditioned behaviors depending on social class, religious, political, and financial relationships.

An additional question also reviewed by de Waal (2005) refers to the behavioral construction of our species members. In other words, do we represent the consolidation of given behavioral phenotypes? Or does our species involve genetically driven interactions of two behaviorally different subspecies—such as those represented by the chimpanzee and the bonobo main social, behavioral profiles, as mentioned before? Are we chained to a behavioral bipolarity by which we condemn and at the same time exercise cruel behaviors against supposedly menacing or culturally "different" conspecifics? Do our genes carry both profiles, expressed differently within the human population? Could it also be expressed in terms of profit, prevailing, or dominance-prone versus solidarity or social sensitive-prone individuals? Furthermore, are we at the evolutive end of the primate order? Will time and technological-driven human evolution allow for a different *Homo* evolutionary path as it has done throughout evolution? Moreover, do those possibilities contradict the concept of universals of animal behavior as chained to natural evolution of primate species?

Mistic and religious drives have emerged since the early times of our species evolution. From one point of view, they will endure if fear dominates our frail sense of physical survival in a menacing world. Would it survive as humans progress in their quest for a rational understanding of life? Would new emotional and rational constructions conquer the very basal neurobehavioral primeval, fear-fragile basic plan anchored in the configuration of basal brain neural circuits? In due time could our species—or a derived subspecies—rewire the involved circuitries that keep us embraced to our ancient animal evolution profile, based on mere survival and prevalence for survival as the concept of selfish genes suggests? Are they armored against or foreign to new circuital behavioral influences stemming from higher—neocortical related—behavioral modes?

Several of these questions could generate a wealth of hypothesis. Perhaps they could start to be answered once societies foster merit, equal opportunities, and change cultural values toward social recognition and praise, as opposed to material profit and its spin-off consequences of dominance and social privileges.

In a world with severe extremes, answering these questions have the scent of a utopian character. However, the answers we give to them are the ones that could open new evolutionary roads and social construction.

The other dimension involves the impact of human activities on climate and ecological degradation, poverty, and social and political unrest. These domains have been a matter of public discussion fueled by several international organizations and reviewed in previous publications (Colombo 2010, 2015, 2019, 2021) and summarized in the following excerpts:

> Globally, about 25 percent of the total land area has been degraded... Scientists recently warned that 24 billion tons of fertile soil was being lost per year, largely due to unsustainable agriculture practices. If this trend continues, 95 percent of the Earth's land areas could become degraded by 2050. Globally, 3.2 billion people are affected by land degradation,

especially rural communities, smallholder farmers, and the very poor.[1] (GEF, Global Environmental Facility, 2019)

Across the world, climate change, the loss of biodiversity, land degradation and water scarcity are growing problems that need to be urgently addressed if the world is to achieve the goals set out in the 2030 Agenda for Sustainable Development, the reports state.[2] (Global Environmental Outlook (GEO-6), 2016)

Today, humans extract more from the earth than ever before (~60 billion tons of renewable and non-renewable resources) {2.1.2} with population doubling over 50 years {2.1.4} and the per person consumption of materials up 15% since 1980. Since 1970, global extraction of biomass, fossil fuels, minerals, and metals increased sixfold {2.1.6, 2.1.11, 2.1.14}. Urban area doubled since 1992 and half of agricultural expansion (1980–2000) was into tropical forests {2.1.13}. Fishing now covers over half the ocean {2.1.11}. Since 1980, greenhouse gas emissions doubled {2.1.11, 2.1.12}, raising average global temperature by at least 0.7 degrees {2.1.12} and plastic pollution increased tenfold {2.1.15}. Over 80% of global wastewater is discharged into the environment without treatment, while 300–400 million tons of heavy metals, solvents, toxic sludge, and other wastes are dumped into the world's waters each year {2.1.15}. Fertilizers enter coastal ecosystems, producing more than 400 hypoxic zones and affecting a total area of more than 245,000 km {2.1.15}. The number of recorded invasive alien species doubled over 50 years {2.1.16}. Today, a full 75% of the terrestrial environment, 40% of the marine environment, and 50% of streams manifest severe impacts of degradation {2.1.12}.[3] (Intergovernmental Science-Policy Platform on Biodiversity and Ecosystem Services (IPBES), 2019)

According to the Intergovernmental Panel on Climate Change, UN,[4] climate change will not be uniform and proportional to the potential levels of global warming (expected to be between +1.5 °C and +3.0 °C) and will have different impacts on different parts of the globe.

The National Geographic comment summarized the late IPCC report on climate change:

The 'code red alert' for the planet, issued Monday by the United Nations' Intergovernmental Panel on Climate Change, is an unsurprising, if unsettling message. It means there's no going back—the warming of the planet will intensify and keep pushing key climate systems toward irreversible change. The latest report, the sixth the IPCC has published since 1990, firms up the connection between global warming and extreme weather.[5]

As mentioned earlier, the 2020 World Bank Group Report states:

About a quarter of the global population is living below the US$3.20 poverty line, and almost half is living below the US$5.50 line, compared with less than a 10th living below US$1.90. These figures translate to 1.8 billion people and 3.3 billion people at the US$3.20

[1] https://www.thegef.org/topics/land-degradation

[2] https://www.un.org/sustainabledevelopment/blog/2016/05/rate-of-environmental-damage-increasing-across-planet-but-still-time-to-reverse-worst-impacts/

[3] https://zenodo.org/record/5018938#.YRGKpVMzaJE

[4] https://www.ipcc.ch/assessment-report/ar6/

[5] https://www.nationalgeographic.com/environment/article/some-irreversible-changes-to-the-climate-can-still-be-headed-off-report-says?cmpid=org=ngp::mc=crm-email::src=ngp::cmp=editorial::add=Planet_Possible_20210810&rid=95527DDF86250D8AB073474627963400

> and US$5.50 poverty lines, respectively. The number of people living below US$3.20 today is as high as the number of people in extreme poverty in 1990, the starting point of this analysis, which is perhaps one way to illustrate the scale of the challenge that remains at these higher lines. The number of people living below US$5.50 per person per day has barely declined over the past 25 years.

Additionally, according to UNICEF (2021):[6]

> Billions of people will lack access to safe water, sanitation and hygiene in 2030 unless progress quadruples – warn WHO, UNICEF.

And the Intergovernmental Panel on Climate Change (2018):[7]

> Global warming of 1.5 °C.
>
> An IPCC Special Report on the impacts of global warming of 1.5 °C above pre-industrial levels and related global greenhouse gas emission pathways, in the context of strengthening the global response to the threat of climate change, sustainable development, and efforts to eradicate poverty. (2018 Intergovernmental Panel on Climate Change)

According to data reported in 2021 by Our World in Data,[8] three billion people cannot afford a healthy diet (based on the concept that a wide range of nutrient-dense foods are needed to obtain the required vitamins and minerals). The following partial listing includes countries with a percent of the population over 70% and below 2% that cannot afford a healthy diet:

Over 70%

Angola, Bangladesh, Benin, Botswana, Burkina Faso, Central African Republic, Chad, Congo, Democratic Republic of Congo, Egypt, Ethiopia, Gambia, Guinea, Haiti, India, Kenya, Laos, Liberia, Madagascar, Mali, Mozambique, Niger, Pakistan, the Philippines, Rwanda, Senegal, Sierra Leone, Sudan, Tanzania, Togo, Uganda, Zambia

Below 2%

Australia, Austria, Belarus, Belgium, Canada, Chile, Cyprus, Czechia, Denmark, Finland, France, Germany, Iceland, Ireland, Israel, Japan, Malta, the Netherlands, Norway, Poland, Portugal, Spain, Switzerland, the United States, Uruguay

From the same source (report from *Our World in Data*), energy consumption per person/per country reveals extreme differences among them, most especially those with consumption at or below 3.5 KW (considered as extreme energetic poverty), the threshold consumption for minimal living requirements. The same source notes that the 20% most rich worldwide consumes 80% of all energy requirements.

These reports stress a conflicting concept amid extreme inequalities in living conditions and resources, that is, whether human development progresses on unsustainable grounds that affect our ecosystem and climate or can be managed with

[6] https://www.who.int/news/item/01-07-2021-billions-of-people-will-lack-access-to-safe-water-sanitation-and-hygiene-in-2030-unless-progress-quadruples-warn-who-unicef

[7] 2019 Intergovernmental Panel on Climate Change published Report

[8] H. Ritchie, *A healthy diet: three billion people cannot afford one.* https://ourworldindata.org/diet-affordability

alternative progressive goals, technologies, and ecologically safe exploitation and consumer culture—accepting the need for a different developmental pace and assuring it reaches populations worldwide. This statement contrasts with the consequences of human actions based on free-running political and corporative interests and an implicit dominance/prevalence behavior. They result in profoundly unequal living quality and reduced opportunities to avoid the consumerist pressure, with added climate and menacing environmental impacts. Such inequality reminds of ancient, hierarchical animal drives now exceeding the limits of a herd and having worldwide consequences, as if the ancient, primeval, hierarchical access to basic needs has been replaced by more sophisticated means (social standing, buying power, educational cognitive enhancement) to manage—or generate—inequalities in development opportunities and disregard people's equal rights.

Given the world constructed by our species, in which solidarity and creativity tend to be overpowered by financial prevalence and profit-seekers, which in turn control conditions of individual and public cognitive growth, behavioral drives, and consumer habits, a change of social values seems to represent a bleak horizon. However, the very bipolar condition of the species opens a hopeful perspective, involving an internal, continued conflict of values and goals within communities and among individuals. It is a continued cultural battle that could draw divergent paths for the future of our species. A path characterized by the quest of survival (whether physical, emotional, social, or cultural in its various domains) fueled by individual and group dominance behavior.

Regardless of our extraterrestrial visions of conquest—and their corporative promoters—as members of the worldwide village of our species, it must be made conscient that we cannot survive with human dignity unless the continued expression and prevalence of ancient animal drives are contained and replaced by humane values and behavioral profiles that promote survival, cooperation, and solidarity within and among its various profiles of cultural expressions.

There is such an astounding amount of money invested in warfare, extraterrestrial projects, and political corruption (Colombo 2015, 2021) that could be applied to foster basic life quality, cognitive enhancement, public health and environmental protection, and research on alternative energy sources.

Despite the Space Treaty signed in 1967—according to which outer space, including the moon and other celestial bodies, may not be nationally appropriated by claiming sovereignty, use or occupation, or in any other way—China, the United States, Russia, Israel, and Japan have launched a race toward extraterrestrial targets, as private corporate interests have also done. Does UN regulations also apply to corporations or individuals, whether associated or not to federal institutions?

A few examples of gross money resources applied to extraterrestrial exploration provide an overview of this issue (in addition to warfare expenses, as mentioned in previous chapters):

- Over the last 3 years, governments spent a combined total of USD 216.27 billion on space activities. The budget approved for NASA for 2021 is USD 23.3 billion, an increase of 3% from 2020.[9]

[9] https://africanews.space/global-space-budgets-a-country-level-analysis/

- Since its inception, the United States has spent nearly US$650 billion (in *nominal dollars*) on NASA.[10]
- Fifty years after the Apollo 11 astronauts first walked on the moon, global investment in space is blasting off again. According to *new data* from space-market specialists Euroconsult, last year, 18 countries spent at least $200 billion on space-related activities, from satellites to exploration. While the United States remains on top, other nations are gaining. China has increased space spending 349% over 15 years. That's just on the government side.
- Private investment in space is also reaching a new orbit. A total of $22.3 billion has been invested in 476 space companies since 2009, according to investment firm Space Angels.[11]
- The United States spent more than $200 billion on the space shuttle and another $50 billion on the International Space Station. From its creation in 1958 through 2018, the National Aeronautics and Space Administration (NASA) spent almost one trillion inflation-adjusted dollars.[12]
- According to the Stockholm International Peace Research Institute, in 2014, roughly US$1.8 trillion was committed to military expenditures worldwide.[13]
- As stated by Peter Diamand, in the 2006 International Space Development Conference, "There are twenty trillion dollar checks up there, waiting to be cashed!"[14]

Among other promoters, extraterrestrial exploration is fueled by similar motivations to those operating at ground base Earth: relative power stance on dominance (in terms of politics and military strategy strength), mining profit, and territory occupation rights. In summary, the quest for dominance overrides potentiating technological progress applied to common needs and improving our standing with ecological protection.

On the optimist side, we have sporadic lively, individual, and collective expressions involving social and ecological concern, creativity, and solidarity and technological improvements affecting various domains of quality of life in population conglomerates.

Could our species behavior construed on ancient, neurologically entrenched, animal drives continue to succeed in increasing basic survival and dominance drives—fueled by profit/advantage seekers—or rather increase people's cognitive awareness and avoid the menace to the viability of our complex, dynamic, ecological niche?

Would dominance drive compromise the future of human cooperation and replace common welfare and growth by committing their goals to *lukrum* or to for-profit enterprises blind to ecology and common development and well-being? Would unyielding profit-seeking behavior—sequelae of dominant behavioral

[10] https://en.wikipedia.org/wiki/Budget_of_NASA

[11] https://fortune.com/longform/space-program-spending-by-country/

[12] https://billofrightsinstitute.org/activities/was-federal-spending-on-the-space-race-justified

[13] https://interestingengineering.com/is-it-worth-it-the-costs-and-benefits-of-space-exploration

[14] https://space.nss.org/book-review-mining-the-sky/

trends—drive human societies to further survival risks than those evolving in modern times? Must human cognitive evolution place at risk the very basic grounds of our species survival and foster social inequalities? In such case, would our species fall into an *evolutive trap*, where its own comparative advantages would drive into species isolation and exhaust ecological survival? To what extent would unchecked quest for profit and technological advances consolidate a dominant-prone species and technological-based castes within and beyond Earth bounds? In this case, are we heading into a deepening of sociocultural gap and the construction of a "eusocial" organization comparable to the one described in insect species with fixed hierarchies and social roles?

Among these hypothetical variants, there should be an alternative for sociality based on solidarity, equal rights, and opportunity for collective development. Although there is a long way to utopia, our species should not abandon such a chimeric horizon (artistic interpretation in Fig. 13.2).

Fig. 13.2 Fine arts interpretation of "A long way to *Utopia*" 2014. Mixed technique, 120 × 80 cm. (Author J.A.C.M.)

References

Colombo JA. Somos la especie equivocada? (Are we the wrong species?). Ed. EUDEBA, Buenos Aires, 2010.

Colombo JA. Los Homo sabios (The wise Homo), Ed. Buenos Aires Books, Buenos Aires, 2015.

Colombo JA. Our Animal condition and social construction. 2019. Nova Sci. Publ., USA. ISBN: 978-1-53615-357-6. eBook ISBN: 978-1-53615-3583.

Colombo JA. The Homo within the sapiens. 2021. Nova Sci. Publ. USA.

Dawkins R. The selfish gene. Oxford University Press; 1995.

de Waal F. Our inner Ape. Penguin Group: Berkeley Publishing Group; 2005.

IPBES (2019): Global assessment report on biodiversity and ecosystem services of the Intergovernmental Science-Policy Platform on Biodiversity and Ecosystem Services. E. S. https://doi.org/10.5281/zenodo.3831673.

Lin D, Hanscom L, Murthy A, Galli A, Evans M, Neill E, Mancini MS, Martindill J, Medouar F-Z, Huang S, Wackernagel M. Ecological Footprint Accounting for Countries: Updates and Results of the National Footprint Accounts, 2012–2018. Resources 2018, 7, 58; https://doi.org/10.3390/resources7030058.

Kitzes J, Wackernagel M, Loh J, Peller A, Goldfinger S, Cheng D, Tea K. Shrink and share: humanity's present and future Ecological Footprint. Philos Trans R Soc (B). 2008;363:467–75.

World Bank Group. "Poverty and Shared Prosperity 2020. Reversals of Fortune", page 39.

Printed by Printforce, the Netherlands